Gravitational Renaissance:

Unveiling the Universe Through Evolving Compressive Gravity

Dr. James W. Michel

Cosmic Chronicles Monograph 1

Title: Gravitational Renaissance: Unveiling the Universe Through Evolving Compressive Gravity. Cosmic Chronicles Monograph 1

Author: Dr. James W. Michel

Publisher: Cosmic Chronicles Creative: Educational Publishing for Children and Adults

Library of Congress Control Number: 2024920167 **ISSN:** 30650488

ISBN Number: ISBN-13: 978-1-965466-00-1 (Paperback)

Edition: First Edition. **Printing:** Printed in the United States of America

Publisher's Address: CCC Pubs, 1122 NW 10th Ave., Camas WA 98607

Contact Information: URL: https://www.doctorjamesmichel.com

Email: support@cccpubs.com Text Line: +1 (831) 869-6821

Endorsements:

Dedication:

To the powerful women in my life—my mother, grandmothers, great grandmother, sisters, wives, and my life partner—each of whom has infused my existence with greater humanity through their unique gifts and profound influence.

Table of Contents

Genesis of a New Theory

This monograph represents not just the culmination of a lifelong fascination and dedicated inquiry into the fundamental forces that orchestrate the cosmos but also a tapestry woven from threads of ancient wisdom and modern understanding. These threads stretch back to conversations—both real and envisioned—with giants of science such as Sir Isaac Newton, whose laws of motion set the stage for all classical mechanics; Albert Einstein, who redefined gravity with his theory of general relativity; and Stephen Hawking, who explored the very boundaries of black holes and the universe.

My journey into the depths of gravitational theory began in childhood, guided by my father, Dr. Lester Michel, during our many climbing trips in the Colorado mountains. He introduced me to the wonders of the physical world through demonstrations and discussions that sparked my curiosity and challenged my understanding. These cherished moments laid the groundwork for my lifelong quest to understand the forces that govern our universe.

As I delved deeper into the realms of physics and cosmology, it felt as though I was collaborating with the spirits of these great scientific minds. In moments of deep reflection and study, I experienced profound insights

that seemed as if directly conveyed by these historical figures, suggesting a narrative of discovery that was not mine alone but shared with those who laid the foundational stones of our understanding of the universe.

In this monograph, I propose a new theory—Evolving Compressive Gravity—which radically reimagines the force of gravity. Casting aside the traditional view of gravity as a pulling attraction between masses, this theory presents it instead as an emergent, omnidirectional pressure, postulated as a manifestation of the universe's acceleration in a hypothetical fourth dimension—a speculative but potentially enlightening perspective.

The genesis of this theory was not a sudden revelation but a gradual illumination, piecing together a cosmic puzzle where each piece was handed down through generations of tireless seekers. In sharing this journey, I aim not only to present a new perspective on gravity but to honor the legacy of those luminaries who have illuminated our path.

Thus, we begin at the beginning—at the first inklings of curiosity that sparked an expedition across centuries and consciousness, to explore the most profound questions of our existence. This exploration serves as a precursor to a deeper journey towards understanding the interconnectedness of all physical laws, a quest that I hope will lead us closer to a unified

theory of everything. As we embark on this voyage through the monograph, each chapter will build upon the last, exploring the implications of this theory from cosmic phenomena to the potential for future technological advances.

Chapter 1.1: Historical Perspectives on Gravity

Section 1.1.1: Gravitational Concepts Through the Ages

Purpose of This Exploration:

Across the woven tapestry of time, the quest to understand gravity has been a pursuit of the human spirit's deepest curiosities. This section traces the evolution of gravitational thought, from the philosophies carved in the ancient stones of civilization to the celestial calculations that defined the Renaissance skies. In retracing these steps, we begin to unravel the grand mystery of the cosmos—a journey not merely of scientific endeavor but of profound transformation in how humanity perceives its universe.

Significance of Historical Context:

Gravity, that subtle whisper of the cosmos, has long danced through the human mind, guiding planets and pulling apples from their branches. It has been a muse to poets and a forge for the philosophies that shaped our intellectual heritage. Each epoch's grasp of gravity has mirrored its own understanding of the cosmos, reflecting broader narratives about life, existence, and the structure of reality. As we embark on this historical voyage, we set the stage for a revelation, preparing to shift paradigms with a new theory as audacious as it is illuminative.

The Ancient Foundations

In the shadowed halls of antiquity, where philosophers pondered the

order of the cosmos, gravity was conceived as a divine choreographer, orchestrating the celestial and the terrestrial in a harmonious ballet. The ancients asked not only what the world was made of but where it naturally sought to be.

Aristotle's Natural Order:

For Aristotle, the universe was an arena of purpose and place. Objects, driven by their inner nature, sought their rightful position in a cosmic order. Earth sought the center; air and fire, the heavens. This theory, a celestial symphony of simplicity and elegance, dominated the stage of the medieval mind, unwavering through the ages.

The Renaissance Shift

As the Renaissance dawned, the skies began to whisper secrets to those who would listen. The rebirth of science brought with it a telescope, not just of glass and metal, but of thought and perspective. The heavens were no longer a static realm of crystalline spheres but a dynamic universe of suns and planets, bound by unseen forces yet to be understood.

Copernican Revolution:

Nicolaus Copernicus, with a bold stroke of the quill, recentered the cosmos, placing the Sun, not the Earth, at its heart. This idea, radical and redefining, cracked the ancient foundations, challenging scholars to rethink not just the stars but the forces that guided them.

Galileo's Insights:

Galileo Galilei, the audacious observer of skies, shattered the Aristotelian

doctrine with simple yet profound experiments. His insight that all objects, regardless of weight, shared the same fierce desire to fall, spoke of a universe more intricate and wonderful than previously imagined. His telescopic discoveries declared that the universe was larger, more marvelous, and more mystifying than ever conceived.

Opening Insights

This opening narrative is not merely about the evolution of ideas; it is a chronicle of humanity's relentless pursuit to understand the force that holds the stars and us in its embrace. As we peel back the layers of history, each epoch's insights build towards a crescendo—a theory that promises not just to redefine gravity but to transform our understanding of the cosmos itself.

Section 1.1.2: Building Foundations in Ancient Wisdom and Renaissance Insight

Echoes of Ancient Theories

Before telescopes and treatises, ancient civilizations grappled with the mysteries of motion and force. These inquiries formed the foundational stones of scientific thought, which would later be built upon by the architects of the Renaissance.

Aristotle's Cosmic Vision

In ancient Greece, Aristotle envisioned the universe as a realm of order, where celestial bodies moved according to divine guidance and objects on Earth sought their natural place. His theory, integrating elements with motion, dominated for centuries, illustrating how our ancestors sought not only to understand the cosmos but to explain its perfect order.

Medieval Cosmological Tapestry

Medieval scholars, bridging the ancient to the Renaissance, intertwined Greek philosophy with the celestial hierarchy of Christian theology, creating a grand cosmological tapestry. This period reinforced the geocentric view and Aristotle's theories, setting the stage for a dramatic paradigm shift during the Renaissance.

Dawn of the Renaissance

The Renaissance, a period marked by a rebirth of science and culture, saw a resurgence in the questioning of classical doctrines. This era ignited a transformative shift in understanding the universe, primarily driven by revived interest in empirical evidence and direct observation.

Copernicus Challenges the Cosmos

Nicolaus Copernicus proposed a radical idea: the Sun, not the Earth, was the center of the universe. His heliocentric model redefined celestial mechanics, challenging scholars to rethink not just the stars but the forces that guided them. Copernicus laid down the conceptual groundwork for a universe governed by laws that could be observed, tested, and understood.

Galileo's Empirical Insights

Galileo Galilei, armed with his telescope, embarked on a journey that would challenge the very heavens. Observing Jupiter's moons and the phases of Venus, he provided concrete evidence against the perfect, unchanging spheres of Aristotle's universe. Galileo advocated for a mathematical description of the universe, where celestial bodies followed predictable laws not bound by philosophical ideals but by observable truths.

Bridging Philosophy to Observation

The shift from speculative philosophy to empirical observation marked the true beginning of modern science. The Renaissance set the stage for the Scientific Revolution, transforming gravity from a mystical force aligned with divine intention to a phenomenon that could be systematically and mathematically analyzed. His heliocentric model redefined celestial mechanics, challenging scholars to rethink not just the stars but the forces that guided them. Copernicus laid down the conceptual groundwork for a universe governed by laws that could be observed, tested, and understood.

Section 1.1.3: The Newtonian Revolution

Unveiling the Universal Law

In the twilight of the 17th century, a convergence of celestial and terrestrial insights was crystallized in the mind of Isaac Newton. Amidst the blossoms of an English garden, the simple fall of an apple whispered secrets of the cosmos, sparking a revolution not just in thought but in the very way humanity perceived the universe. Newton's law of universal gravitation emerged as a beacon of enlightenment, casting light on the shadowed mysteries of celestial mechanics with a clarity that was both profound and elegant.

Newton's Synthesis

Isaac Newton, with the publication of his Principia Mathematica, wove a tapestry of the heavens and the earth, bound together by the same invisible threads of attraction. Here was a universe interconnected by a force as applicable to the orbit of the moon as it was to the fall of a petal. Newton proposed that this gravitational pull was universal; every object in the universe reached out to every other with a force that scaled with mass and diminished with distance. This simple yet sublime idea was a symphony of the spheres in its own right, harmonizing the motions of celestial bodies with the mundane meanderings of earthly objects.

Impact on Science and Philosophy

The ripples of Newton's ideas extended far beyond the boundaries of physics, stirring the realms of philosophy and the practical world alike. His laws offered a predictable, measurable universe, operable by knowable laws—accessible through reason and observation rather than metaphysical speculation. This shift empowered a new age of empirical science, where hypotheses could be tested and the cosmos itself became a place of study, not just of wonder. The mechanistic universe Newton painted was like clockwork; intricate, deterministic, and infinitely dissectible in its workings.

Limitations and the Call for Deeper Understanding

Yet, for all its precision and predictive power, Newtonian physics was not without its enigmas. The orbit of Mercury, with its peculiar precession, eluded full explanation under Newton's laws. The nature of gravity itself—how it acted instantaneously across vast distances, what mediated its force—remained a veiled mystery that Newton himself confessed was an "absurdity" he could not yet resolve. These puzzles laid down the gauntlet for future physicists, hinting at deeper truths veiled beneath the surface of Newtonian mechanics.

Newton's Enduring Legacy

Newton's legacy is a testament to the power of a unifying idea. His principles underpin much of modern technology, from the trajectories of spacecraft to the calculations that keep bridges standing and GPS satellites

orbiting. Yet, his greatest contribution may well be the framework he provided for asking better questions—questions that would one day require new physics for their answers. His work set the stage for the giants who would follow, those like Einstein, who would stand on Newton's shoulders to gaze even deeper into the universe's mysteries.

Section 1.1.4: Einstein's Relativity

Unraveling the Fabric of Cosmos

At the dawn of the 20th century, the stage was set for a revolution in our understanding of the universe. Albert Einstein, a thinker of audacious insight, prepared to challenge the very fabric of cosmic understanding. With relativity, he did not merely reformulate the laws of gravity; he reimagined the theater of the universe itself—a cosmos more intricate and intertwined than any had dared to dream.

The Genesis of General Relativity

Einstein's journey through the enigmas of space and time began with an elegantly simple yet profoundly disturbing question: What if light itself was subject to gravity's embrace? This query led to his groundbreaking theory of general relativity, which proposed that the presence of mass and energy could curve the very fabric of spacetime. Here was a universe where gravity was no mere force but a dynamic expression of the geometry of space and time itself.

A New Understanding of Gravity

In Einstein's universe, gravity emerged not as a force that acted, but as a consequence of the bent paths through a curved spacetime. This radical idea explained why planets orbited stars and why light from distant stars bent around our sun. It offered solutions to puzzles that had perplexed astronomers under Newtonian physics, such as the peculiar orbit of Mercury. Einstein's equations suggested that gravity was the music played on

the keys of spacetime, a symphony of the cosmos that was both elegant and mysterious.

Implications for Cosmology and Beyond

The implications of Einstein's theory stretched far beyond the orbits of planets and the paths of starlight. His vision predicted phenomena such as black holes—regions of space where time and space end—and gravitational waves, ripples in the spacetime fabric caused by cataclysmic cosmic events. These predictions have since been confirmed, transforming theoretical artifacts into observational cornerstones of modern astrophysics.

Challenging the Boundaries of Perception

Einstein's theory pushed the boundaries of human thought, challenging us to imagine a universe where spacetime could stretch, compress, and ripple like a cosmic ocean. His vision showed us that the universe was not a static arena for celestial mechanics but a dynamic, evolving tapestry woven from the geometries of spacetime. His ideas reshaped our understanding of the universe, altering our perceptions of reality itself.

Legacy and Unanswered Questions

While Einstein's theory of relativity remains one of the greatest intellectual achievements in physics, it also highlights the boundaries of our understanding. The theory struggles to reconcile with the quantum world, suggesting that deeper, more fundamental principles remain undiscovered. As such, general relativity is not the final word but a profound chapter in an ongoing saga of discovery.

Section 1.1.5: Modern Challenges

Introduction to Contemporary Gravitational Puzzles

In the vast cosmic theater, where the lights of stars and galaxies play out an age-old drama, modern physics has encountered puzzles that defy easy explanation. These challenges, emerging from the shadows of our understanding, underscore the limitations of our grandest theories—revealing not just gaps in our knowledge but beckoning toward deeper cosmic truths.

The Dark Matter Enigma

Amidst the swirling galaxies, where stars orbit unseen centers and galaxy clusters bind themselves in silent fellowship, a mystery lurks—dark matter. This unseen substance, inferred not by the light it emits but by the gravitational grip it exerts, puzzles astronomers and physicists alike. Its presence is a whisper in the cosmic winds, a ghostly entity that holds over eighty percent of the universe's mass yet eludes direct detection. Dark matter, a concept as elusive as it is essential, challenges our understanding of the universe's very structure.

The Accelerating Universe and Dark Energy

A revelation came from the cosmos in the late 20th century, as distant supernovae told tales of an expanding universe accelerating under an unseen force. This discovery, as startling as it was profound, suggested the presence of dark energy—a mysterious force pushing the cosmos apart, overwhelming the gravitational pull that seeks to slow it down. Comprising about seventy percent of the universe's energy density, dark energy remains one of the great conundrums of modern cosmology, a riddle written across the very fabric of spacetime. The James Webb Space Telescope (JWST) has added complexity to this narrative by observing galaxies far beyond the expected cosmological horizon, suggesting that the universe's expansion might not only be accelerated by dark energy but also more variable and extensive than previously thought.

Quantum Gravity and the Unification Challenge

At the smallest scales of existence, where the laws of quantum mechanics reign, gravity refuses to conform to expected rules. The quest for quantum gravity—to marry the macroscopic rules of general relativity with the microscopic laws of quantum physics—remains one of the most profound challenges in theoretical physics. Attempts like string theory and loop quantum gravity have offered glimpses into possible unifications, yet a complete and consistent theory eludes our grasp. This ongoing struggle not only highlights the limitations of our current frameworks but also illuminates the path toward potential revolutionary insights.

Legacy and the Horizon of Discovery

The legacy of these modern challenges is not merely in their questions but in the doors they open for future explorations. As we stand on the precipice of knowledge, looking out over the landscape of the unknown, these puzzles invite us to rethink, reimagine, and reconstruct our understanding of the universe. It is here, at the edge of mystery, that new theories like Evolving Compressive Gravity find their call to action—a promise to illuminate the dark and chart a course through the unexplored.

Section 1.1.6: Introducing Evolving Compressive Gravity

Introducing Evolving Compressive Gravity

In the grand tapestry of the cosmos, where each thread of understanding weaves into the next, the theory of Evolving Compressive Gravity emerges as a bold new pattern, reshaping our perception of the universe. This theory does not merely adjust the existing framework but rethinks gravity's very essence, proposing it as an emergent, not fundamental, force—a product of the universe's expansion into unseen dimensions.

Rethinking Gravity as an Emergent Phenomenon

Gravity, as traditionally understood, pulls apples to earth and binds galaxies together. Yet, under Evolving Compressive Gravity, this force is reimagined as a dynamic pressure exerted by the fabric of the cosmos itself as it expands through dimensions beyond our perception. This paradigm shifts our understanding from a universe where gravity is a pulling force to one where it is a pressing reality from all directions, an echo of the universe's continual expansion.

Addressing Modern Cosmological Challenges

This new vision of gravity offers elegant solutions to some of modern physics' most persistent puzzles. It eliminates the need for dark matter by explaining galactic curves with its innovative gravitational dynamics. Similarly, it replaces dark energy with a natural property of spacetime's geometry, providing a compelling explanation for the accelerating expansion of the universe.

Theoretical Foundations and Mathematical Formulation

Evolving Compressive Gravity is grounded in a robust mathematical framework that extends the equations of general relativity to incorporate the effects of a fourth spatial dimension. These equations do not contradict Einstein's findings; rather, they build upon them, introducing new variables that account for the additional forces at play.

Implications for Physics and Beyond

The implications of Evolving Compressive Gravity are profound, extending beyond astrophysics to influence fields such as quantum mechanics, cosmology, and even philosophical inquiries into the nature of reality. By redefining gravity, this theory invites us to reconsider everything from the movement of particles to the expansion of the cosmos, opening new avenues for technological advancement and scientific exploration.

Section 1.1.7: Legacy and Unanswered Questions

The Enduring Influence of Einstein's Relativity

Albert Einstein's insights into the curvature of spacetime have echoed through a century, fundamentally altering every facet of our understanding of physics. The legacy of his theory is vast, stretching from the blackboards of theoretical physicists to the operational systems of satellites that navigate the globe. Each validation of relativity, from the bending of starlight to the direct detection of gravitational waves, has reaffirmed its foundational role in modern science.

Resolving and Revealing Cosmic Mysteries

Relativity has not only resolved mysteries but also revealed new ones. Its success in describing large-scale cosmic phenomena is unparalleled, yet it paradoxically highlights profound gaps in our understanding at quantum scales. This juxtaposition of triumph and limitation exemplifies the dynamic nature of scientific inquiry—where answers often lead to more profound questions.

Unresolved Issues and the Need for New Theories

While Einstein's framework predicts the behavior of stars and galaxies, it falters when faced with the quantum world. The elusive graviton, the theoretical quantum of gravity, remains undetected, and the mathematical integration of gravity with quantum mechanics remains incomplete. These gaps signify the unfinished business of theoretical physics and the fertile ground for new theories.

Setting the Stage for Evolving Compressive Gravity Theory

Einstein's legacy, rich and profound, sets the stage for the next revolutionary theory—Evolving Compressive Gravity Theory (ECGT). This new theory seeks to weave together the disparate threads of quantum mechanics and cosmology into a cohesive fabric that explains not just the structure of the universe but its dynamic, evolving nature. By addressing the unresolved questions left by Einstein's relativity, ECGT promises not only to fill the gaps but to transform our understanding of the universe at its most fundamental level.

Chapter 1.2: Theoretical Foundations and Mathematical Framework of Evolving Compressive Gravity Theory (ECGT)

Section 1.2.1: Foundations of Evolving Compressive Gravity Theory (ECGT)

Beginning Evolving Compressive Gravity Theory (ECGT):

As we journey deeper into the cosmos's secrets, we encounter the necessity for a theory that transcends traditional boundaries and addresses the universe's enigmas more holistically. ECGT emerges as a beacon in this scientific odyssey, challenging our conventions and proposing a radical new perspective on gravity.

Introducing the Theory:

Recapping our voyage through the scientific legacy of Newton and Einstein, we find that their monumental contributions, while groundbreaking, left us with puzzles that linger on the fringes of our understanding. ECGT arises in response to these challenges, weaving the observed

phenomena into a coherent theory that redefines gravity's very essence. It posits that gravity is not a pulling force but an emergent, dynamic interaction within a universe that is richer and more complex than the three-dimensional space we experience.

Core Principles of the Theory:

Central to ECGT is the principle that gravity is an emergent phenomenon—a byproduct of interactions within a fourth spatial dimension. This dimension, unseen yet profound in its implications, adds a layer of depth to the cosmic fabric. Here, gravitational effects emerge completely independently from the intrinsic properties of mass, arising strictly from the dynamic pressures exerted by the expansion of the universe into this additional dimension. Mass acts to absorb this pressure, shielding it from nearby masses.

Theoretical Justifications:

This theory elegantly addresses the dark matter conundrum by proposing that what we perceive as the gravitational pull from unseen mass is actually the influence of compressive forces from higher-dimensional space. Similarly, the mysterious acceleration of the universe's expansion, attributed to dark energy, is explained as a natural consequence of the

dynamics within this fourth dimension. Such perspectives not only solve existing puzzles but also prompt a profound shift in how we conceive the universe's structure.

Preliminary Mathematical Overview:

Incorporating the concept of evolving gravitational strength, we adapt the traditional metric tensor of spacetime, expanding it to include varying temporal components that account for the evolving nature of gravitational forces. These adaptations are detailed in Appendix A, where we expand on the modified Einstein field equations to incorporate these new dimensions of analysis. For a detailed derivation, see Appendix A.

Further Reflections and Implications:

Thus, the foundations of ECGT offer a fresh lens through which to view the cosmos—an invitation to rethink not just gravity but the very fabric of reality. As we proceed, these principles will be mathematically articulated and rigorously tested, challenging us to not only understand the universe but to reimagine it.

Section 1.2.2: Mathematical Formulation of the Theory

The Language of the Cosmos: Unveiling New Equations

In our quest to understand the universe through the lens of Evolving Compressive Gravity Theory (ECGT), it becomes crucial to articulate this understanding in the universal language of mathematics. This section unfolds the mathematical tapestry that supports our theory, providing both a solid foundation and profound implications for the nature of gravity.

Shifting to a Fourth Spatial Dimension

The journey into the mathematical heart of ECGT begins with a shift from the traditional three-dimensional view of the universe to a model enriched by a fourth spatial dimension. This additional dimension introduces new dynamics into the gravitational equations, necessitating a novel set of mathematical tools and concepts. These tools allow us to explore how fourth-dimensional influences reshape our understanding of gravitational interactions.

Deriving the Key Equations

Starting with the elegant framework of general relativity, we extend Einstein's field equations to include the influence of the fourth spatial dimension. These extensions do not merely add terms to the equations but rethink how spacetime itself is structured. New variables account for the pressures and densities associated with the fourth-dimensional expansion, and changes in gravitational strength over time. These insights provide fresh perspectives on the gravitational force as an emergent phenomenon and are elaborated upon in Appendix 1.B, where detailed derivations offer a deeper dive into these transformative concepts.

Contrasting Classical and Modern Understandings

By comparing these new equations to the classical laws of Newton and the relativistic equations of Einstein, we illuminate significant divergences and enhancements. The way mass influences gravity in our theory—through its absorptive shielding effect—offers a different perspective on gravitational interaction. This perspective is more aligned with observed galactic behaviors that defy traditional explanations and is rooted in the dynamics of a four-dimensional framework.

Mathematical Integrity and Future Directions

The mathematical formulation of ECGT is crafted to be as rigorous as it is revolutionary. We ensure that these new equations are consistent within the established framework of physics and robust enough to withstand predictive testing. This section addresses potential mathematical challenges and outlines how ongoing research aims to refine and validate our theoretical predictions, further enhancing our understanding of the universe.

Implications and Prospects

The mathematical journey through ECGT deepens our understanding of the universe and challenges us to think beyond the confines of traditional physics. As we continue to explore and test these equations, we move closer to a comprehensive theory that not only explains the known but also illuminates the unknown, promising to reshape our cosmic perspective.

Section 1.2.3: Integration with Existing Theoretical Frameworks

Harmonizing with General Relativity and Newtonian Mechanics

Evolving Compressive Gravity Theory (ECGT) and Classical Theories: ECGT not only reveres but also expands upon the monumental legacies of Newton and Einstein. Despite Einstein's reservations about the scope of Newtonian mechanics, his theories of relativity have profoundly reshaped our comprehension of gravity and spacetime. ECGT emerges from the confluence of these intellectual giants, integrating Newtonian formulas for acceleration with Einstein's principles of spacetime curvature into a vibrant tapestry of four-dimensional interactions. This synthesis broadens both frameworks, offering novel insights into gravitational interactions and addressing enigmatic cosmic phenomena that have long defied explanation, such as the expanded universe observations by JWST.

Revisiting Black Holes and Singularities

A New Interpretation of Black Holes: Traditionally seen as masses with gravitational pulls so intense that not even light can escape, black holes in ECGT are realms where massive objects nearly completely block gravitational pressure—a celestial abyss that compels inward movement not by attraction but by the intense push of surrounding energies and matter. This reinterpretation challenges the traditional singularity paradigm, proposing that these formidable entities are rather distinct points where four-dimensional dynamics intersect visibly with our observable universe, potentially aligning with recent findings that question the uniformity of cosmic expansion.

Discovering White Holes

Theoretical Extensions to Cosmology: ECGT extends the cosmic narrative to include white holes—conjectural counterparts to black holes that, while unobserved, might act as cosmic generators, dispersing matter, energy, and gravitational influences, potentially crafting the universe's architecture from realms beyond our current understanding. This concept, inspired by equations mirroring those of black holes, suggests a more dynamic universe where such phenomena could play critical roles, as hinted by the JWST's observations of distant galaxies beyond expected horizons.

Implications for the Big Bang Theory and Cosmic Energy Distribution

Redefining Cosmic Origins: ECGT offers a fresh take on the Big Bang theory, suggesting a universe sculpted by phenomena akin to white holes, which provide a palpable and relatable origin story free from the enigmatic singularities and exotic conditions typically postulated. This perspective aligns with the JWST's observations, challenging traditional views of the universe's uniformity and suggesting a more varied and dynamic early universe.

White Holes and Cosmic Inflation

A New View on Inflation: ECGT posits a compelling explanation for the universe's rapid inflationary epoch, viewed as a period when nascent gravitational forces were still evolving, allowing for expansive cosmic growth. This theory speculates a complex interplay between the devouring nature of black holes and the generative potential of white holes, suggesting a grand cosmic ballet that cycles matter, energy, and information across hidden dimensions of space and time, a concept that could reshape our understanding of cosmic inflation and its implications on observable phenomena.

Section 1.2.4: Practical Implications and Future Applications

Solving Puzzles in Astrophysics

A number of mysteries continue vexing astrophysicists. Here we look at some of them:

Galactic Rotation Curves:

One of the most compelling validations for any gravitational theory is its ability to elucidate the flat rotation curves of galaxies. Traditional models require dark matter to explain why stars at the edges of galaxies rotate at the same velocity as those nearer the center. ECGT offers a groundbreaking perspective, suggesting that these phenomena can be attributed to gravitational pressure absorption aligned with the axis determined by the involved masses, effectively reducing the intervening pressure. This dynamic pushes the masses together by the external pressure generated within our four-dimensional spatial structure, potentially obviating the need for dark matter and fundamentally transforming our understanding of galactic mass distribution and mechanics.

Gravitational Lensing Revisited:

Gravitational lensing, the bending of light by massive objects, is pivotal in astrophysics for mapping the distribution of dark matter. ECGT

revises the conventional understanding of spacetime curvature, proposing that light is pressed around massive objects due to their shielding effect, which blocks gravitational pressure in the direction of the mass. These theoretical adjustments could revolutionize methodologies for observing the universe and interpreting gravitational lensing effects, enhancing our comprehension of mass and energy distribution across the cosmos.

Implications for Cosmology

Structure Formation in the Early Universe:

Current theories heavily depend on dark matter to account for the gravitational effects necessary for cosmic structure formation. ECGT disrupts this paradigm by introducing alternative mechanisms such as emergence and an evolving, variably increasing gravitational strength, driven by the dynamics of a four-dimensional acceleration paradigm. This novel approach could elucidate the formation of cosmic structures from the largest galactic clusters to the smallest protogalaxies without relying on mysterious dark matter.

Cosmological Inflation:

ECGT provides a fresh framework for interpreting the rapid expansion of the early universe, traditionally explained by inflationary theory. By incorporating the evolutionary dynamics of a four-dimensional spacetime directly into models of the universe's expansion, including solving for a

variable gravitational strength, ECGT offers a more direct explanation for inflation. This framework suggests that initially weak gravitational forces allowed for rapid expansion of the universe, minimizing the reliance on speculative inflationary particles or fields and aligning cosmic history with a continuous, naturally observable expansion process.

Future Research and Technology

New Predictions for Observational Astronomy:

ECGT is ripe with predictions that could be empirically tested with modern astronomical techniques. For instance, the theory anticipates distinct signatures in the cosmic microwave background radiation, which may refine our estimates of the universe's age, offering a more precise timeline of cosmic events. Additionally, ECGT predicts anomalous bending of starlight, deviating from traditional gravitational lensing predictions expected in three-dimensional space plus time models. These phenomena could be observed with the next generation of telescopes and space observatories. Pursuing these experiments is crucial not only for validating ECGT but also for expanding the frontiers of our cosmic understanding. By observing these specific signatures, astronomers can elucidate the dynamics of four-dimensional space, distinct from the conventional three-dimensional space plus time framework, and its implications on a cosmic scale, potentially leading to a significant paradigm shift in astrophysics.

Section 1.2.5: Theoretical and Philosophical Implications

Integration with Quantum Mechanics and Field Theory

Quantum Mechanics:

ECGT revolutionizes our understanding of gravitational interactions at the quantum scale by suggesting that gravity is an emergent, omnidirectional pressure exerted by the universe, modulated by the presence of mass. This perspective indicates that gravity operates under the same principles at both macroscopic and quantum scales, simplifying quantum gravity and showing that our existing knowledge of fundamental forces might be sufficient to develop a unified theory in three-dimensional space.

Field Theory:

The implications of ECGT extend into field theory, suggesting that interactions traditionally described by quantum fields could also be influenced by dynamics in four-dimensional space. This could provide a more holistic understanding of how forces are unified under ECGT, potentially offering a new framework that incorporates gravitational interactions seamlessly with other fundamental forces.

Rethinking Time and Space

Nature of Time:

ECGT challenges and expands our understanding of time by incorporating the dynamics of a four-dimensional space, emphasizing that time is indistinguishable from change, thereby drawing an equivalence to evolution. These insights prompt a reevaluation of how time interacts with space, especially under evolving gravitational conditions, potentially leading to new insights into phenomena such as time dilation and the flow of time near massive objects.

Structure of Space:

By proposing a universe with an intrinsic four-dimensional aspect, ECGT encourages us to rethink our conventional three-dimensional view of space. This theory suggests that what we perceive as the universe might only be a three-dimensional cross-section of a much richer four-dimensional reality, offering new explanations for phenomena like quantum entanglement.

Philosophical Implications

Reality and Perception:

ECGT shifts our cosmic perspective, urging a philosophical reconsideration of how we conceptualize the universe. It challenges us to view gravity not as an intrinsic force of mass but as an emergent property of the universe's structure, fundamentally altering our understanding of reality.

Implications for Cosmology:

By eliminating the need for dark matter and dark energy, ECGT simplifies our cosmological models to intrinsic properties of space itself, enhancing our ability to predict and understand cosmic phenomena without resorting to unobservable entities.

Future Theoretical Developments

Potential New Theories:

ECGT opens the door for new theoretical developments that might explain both observable phenomena and theoretical constructs operating in higher dimensions, potentially leading to a unified understanding of the fundamental forces of nature.

Impact on Other Sciences:

The principles of ECGT could influence other scientific disciplines, such as chemistry and biology, where understanding the fundamental properties of matter and energy is crucial. Insights from four-dimensional dynamics might lead to new ways of exploring chemical reactions or biological processes.

Section 1.2.6: Future Applications and Technological Innovations

Advanced Propulsion Technologies: Space Travel and Gravity Shields

Drawing inspiration from the concept of black holes as regions where compressive gravity is entirely blocked, ECGT suggests the possibility of developing technology to project gravity shields. These devices could harness the universe's omnidirectional gravitational pressure, enabling spacecraft to be pushed effectively towards their destinations by manipulating gravitational fields. Additionally, converting some of the fourth-dimensional acceleration effects into unidirectional acceleration within our three dimensions could provide a new method of propulsion, potentially enabling faster-than-light travel and revolutionizing interstellar exploration.

Energy Generation Through Gravitational Manipulation

The principles of ECGT could lead to innovative methods for energy generation, potentially enabling the conversion of gravitational fields into usable energy. This could provide a sustainable and abundant source of power, influencing everything from large-scale grid power generation to portable energy sources, and potentially revolutionizing how energy is stored and utilized across various sectors.

Advanced Gravitational Wave Detectors: Mapping the Universe

Building on already demonstrated hypersensitive quantum gravity and location sensors, which have piloted transatlantic flights without satellite-based GPS, ECGT could further enhance these technologies. By expanding their use to measure gravitational strength at a distance, we could revolutionize cosmology and navigation, offering more precise and reliable methods for mapping cosmic phenomena and enhancing terrestrial navigation systems.

Quantum Computing and Information Science: Leveraging Quantum Entanglement

ECGT provides a theoretical basis for establishing and maintaining reliable quantum entanglement, potentially based on a deeper relationship with the fourth dimension. This could lead to significant advancements in quantum computing, allowing for the development of quantum engines that could power everything from massive space stations to medical nanobots. The potential for powerful, reliable wireless energy transfer could revolutionize transportation, communications, and numerous other fields.

Medical Technologies: Diagnostic Tools, Treatment Methods, and Nanorobotic Surgery

Leveraging gravitational measurements, technologies inspired by ECGT could lead to more precise diagnostic tools that detect health conditions based on the subtle gravitational fields of the human body. Furthermore, the principles used in quantum engines could be adapted for targeted drug delivery systems or novel therapeutic modalities, utilizing miniaturized devices powered by quantum technology to treat diseases at the cellular level. ECGT opens the door to employing nanorobotic technology for surgery at intracellular levels, allowing for precise manipulations within individual cells such as direct biochemical manipulation, enzyme modifications, and diagnostic analysis of organelles.

Environmental Monitoring and Management: Climate Prediction Models and Resource Management

The advanced understanding of gravitational dynamics provided by ECGT could enhance climate prediction models, offering more accurate forecasts and improved strategies for environmental management. Additionally, new technologies capable of detecting and exploiting resources based on gravitational variations could transform how we manage and utilize natural resources, from water to minerals.

Chapter 1.3: Evolving Compressive Gravity Theory (ECGT) Explained

Section 1.3.1: Overview of Evolving Compressive Gravity Theory (ECGT)

Introduction to the Theory:

In the landscape of modern physics, Evolving Compressive Gravity Theory (ECGT) emerges as a beacon of innovation, redefining gravity not as a mere force that pulls but as an emergent, omnidirectional compressive force shaped by the universe's dance in a fourth spatial dimension. This radical reconceptualization challenges our most fundamental notions of how celestial bodies interact, proposing a universe where gravitational dynamics are woven into the very fabric of spacetime itself.

Distinctive Features:

ECGT departs dramatically from traditional gravitational paradigms by proposing that gravity, rather than being a direct interaction between masses, is the result of higher-dimensional dynamics. This theory synthesizes elements from classical mechanics and quantum theories to provide a comprehensive framework that seamlessly connects the cosmos's vast expanses with the enigmatic quantum world.

Historical Context and Development:

Inspired by the pivotal discoveries of luminaries like Copernicus and Galileo, who revolutionized our understanding of the cosmos through meticulous observation and bold theoretical leaps, ECGT stands poised to catalyze a similar paradigm shift in gravitational theory. Just as the heliocentric model offered a simpler, more elegant description of celestial motions, ECGT simplifies our understanding of gravitational interactions, providing elegant new insights into long-standing cosmic mysteries.

The Shielding Effect and Its Implications:

At the heart of ECGT lies the shielding effect, a novel interpretation of how gravitational forces manifest. Here, gravitational interactions are seen as reductions in a fourth-dimensional pressure force by mass. This innovative concept reinterprets gravitational pull as an external push resulting from decreased pressure in the space between masses. The force exerted is directly proportional to the masses involved and inversely proportional to the square of the distance between them. Moreover, ECGT suggests that gravitational pressure is not an inherent property of mass but emerges from the dynamic, evolving properties of the universe—evolving from minimal initial strength to its present intensity. This perspective provides a plausible, elegant alternative to dark matter in explaining early cosmological phenomena.

Theoretical Significance and Initial Reception:

Since its inception, ECGT has sparked intense discussion and debate within the scientific community. Its fresh approach to explaining gravitational phenomena invites comparisons to historical shifts in scientific thought, challenging established paradigms and promising to redefine our understanding of the universe. Early empirical support and theoretical validation suggest that ECGT might not only hold the key to new scientific truths but also usher in new technological revolutions.

Summary and Transition to Detailed Exploration:

This introduction lays the groundwork for a deeper dive into the nuances of ECGT, setting the stage for an exploration of its mathematical formulations, empirical validations, and broader implications. As we venture further into this theory, we will unravel the intricate tapestry of forces that shape our universe, guided by the light of ECGT's revolutionary insights.

Section 1.3.2: Fundamental Principles of ECGT

The Nature of Gravity as Compressive Force

In the grand tapestry of the cosmos, gravity has traditionally been depicted as a silent maestro, orchestrating the celestial dance with invisible strings of attraction. Evolving Compressive Gravity Theory (ECGT) challenges this classical portrayal, introducing gravity not as a pull but as a push—a compressive force emerging from the unseen dimensions of the universe. This section unfolds the theory's radical premise: that gravity is the manifestation of a pressure exerted by the fourth dimension, shaped not by the attraction of masses but by their ability to shield against this cosmic pressure.

The Role of the Fourth Dimension in Gravity

Imagine a universe more intricate than previously conceived, where an additional dimension—beyond our reach yet fundamental to our reality—guides the forces we feel but cannot see. ECGT posits this fourth dimension as the cradle of gravitational dynamics, influencing the familiar three dimensions in profound ways. Here, we explore how this hidden dimension reshapes our understanding of gravitational forces, turning the mysterious into the understandable and transforming the invisible into observable effects.

Mass and the Shielding Effect

At the core of ECGT lies the shielding effect, a novel concept that redefines the role of mass in gravitational interactions. Mass does not pull but protects; it shields us from the relentless pressure of the fourth dimension. This section delves into how this protective effect creates what we perceive as gravitational forces: the lessened pressure between masses leads to a push towards each other, a cosmic embrace dictated by the laws of shielding. This relationship, still beautifully aligned with the inverse-square law, reveals a universe more coherent and interconnected than ever before.

Mathematical Representation of Compressive Forces

With elegant equations that whisper the secrets of the cosmos, ECGT provides a mathematical language that describes the universe's new rules. This part of the section lays out these formulas, drawing comparisons with the venerable equations of Newton and Einstein, and highlighting how ECGT simplifies and unifies our understanding of gravitational phenomena. It's a narrative where numbers serve not only to calculate but to tell stories of a universe in constant evolution.

Evolutionary Nature of Gravity

Gravity, according to ECGT, is not a constant but a character in flux, evolving as the universe expands and the dynamics of the fourth dimension unfold. This evolutionary perspective offers a fresh lens through which to view cosmic history, from the fiery afterglow of the Big Bang to the majestic swirl of galaxies. We discuss how this concept aligns with observations of the early universe, providing a simpler explanation for cosmic evolution without the need for dark matter.

Testing and Validation

As with any great scientific theory, the proof lies in the testing. This final part outlines the journey ahead for ECGT, from the drawing board to the depths of space, detailing the experiments and observations that could one day elevate the theory from hypothesis to law. It is a call to arms for astronomers, physicists, and cosmologists to look beyond the seen and probe the pressures of the unseen, to validate or challenge the predictions of ECGT.

Section 1.3.3: Mathematical Formulation of ECGT

The Theoretical Canvas of Numbers and Equations

In the realm of physics, where reality meets imagination, the language of mathematics serves not merely to describe but to reveal. Evolving Compressive Gravity Theory (ECGT) unfolds within this domain, where its revolutionary concepts are articulated through a rich tapestry of equations and models. This section weaves the foundational mathematics of ECGT, casting light on its promise to reshape our understanding of the universe's very fabric.

Crafting the Core Equations

At the heart of ECGT lies a bold reconception of gravity, envisioned not as a pull but as a cosmic push, mediated by the unseen tendrils of a fourth dimension.

- **Foundational Assumptions: Our journey begins with a premise as startling as it is profound: the fourth dimension exerts a uniform pressure throughout our familiar three-dimensional universe, decreased by the presence of mass within it. Here, mass does not attract but rather shields, sculpting the gravitational landscape by absorbing this pressure, creating a gradient that results in what we perceive as gravitational forces.**

- **Derivation of Destiny: From this innovative vision, the mathematical soul of ECGT emerges. We derive equations that model how this fourth-dimensional pressure interacts with three-dimensional mass absorption. These equations bridge the gap between theoretical concepts and observable phenomena, linking every cosmic corner, from the microscopic to the majestic. (See Appendix 1.C for detailed derivations)**
- **Universal Language: Presented through the lens of differential and integral calculus, these equations offer a comprehensive view of the universe, providing tools to explore the effects of gravity across different scales and conditions.**

Dialogue with Giants: Newton and Einstein Revisited

ECGT does not stand alone but in conversation with the titans of theory who have preceded it.

- **Newtonian Echoes: We extend Newton's venerable laws, infusing them with a fourth-dimensional perspective. This dialogue offers corrections that illuminate mysteries such as the perplexing rotation curves of galaxies, providing answers without the crutch of dark matter.**

- **Relativistic Harmonies: Aligning with Einstein, ECGT embraces the geometry of spacetime but introduces a variable that evolves—gravity's strength. This evolution offers fresh insights, suggesting a universe more dynamically intertwined than previously imagined.**
- **Resolving the Cosmos's Conundrums: Through its novel framework, ECGT offers resolutions to the dark puzzles of energy and matter, reinterpreting roles and rewriting cosmic rules.**

Predictive Power: Foreseeing the Universe's Secrets

Beyond explaining, ECGT predicts, inviting us to test its truths against the touchstone of reality.

- **Visions of New Realities: The theory forecasts behaviors of gravity under conditions as extreme as the edges of black holes and as elusive as the potential white holes of creation, offering a new lens to view the universe's birth and growth.**
- **Empirical Quests: Proposing experiments from satellite measurements of subtle gravitational anomalies to laser interferometry that might catch the whispers of spacetime's waves, ECGT stands ready for the rigor of reality's judgment.**

Navigating the Mathematical Labyrinth

As with any daring theory, ECGT faces its trials in the proving grounds of mathematics and observation.

- **Exploring Limits and Possibilities: We scrutinize the mathematical underpinnings, searching for weaknesses to strengthen and complexities to clarify, ensuring the theory's robustness and applicability.**
- **Consistency and Verification: Through simulations and proofs, we test ECGT's resilience, preparing it for the ultimate trial by experiment and observation.**

Epilogue: A New Chapter in Physics

This section closes with a reflection on the transformative potential of ECGT's mathematical journey. It beckons future scholars to delve deeper, explore further, and perhaps, in the echoing halls of the universe, to find new truths waiting in the dimensions yet undiscovered.

Section 1.3.4: Comparative Vistas in Gravitational Theories

A Confluence of Thought and Theory

As we traverse the continuum of physics, Evolving Compressive Gravity Theory (ECGT) emerges as a beacon of innovation, challenging and extending the classical and relativistic paradigms of gravity. This section ventures into a comparative analysis, exploring how ECGT not only aligns with but also redefines our understanding of the universe's most elusive force.

Reframing Newtonian Gravity

In the classical arena, Newtonian gravity serves as the cornerstone of our understanding of the cosmos's mechanical nature, explaining the celestial ballet with elegant simplicity. Yet, ECGT takes us deeper, proposing a universe where mass does not pull but shields, transforming gravitational pull into a push driven by differential pressure gradients. This revelation enhances Newton's framework, providing a nuanced explanation for phenomena such as the perplexing rotation of galaxies—offering solutions where dark matter once filled gaps.

Dialogue with Einstein's Legacy

ECGT respectfully nods to Einstein's General Relativity, a theory that redefined gravity as the curvature of spacetime itself. Building upon this profound insight, ECGT introduces an evolving dimension of gravity, adding a dynamic quality to the spacetime fabric. This approach addresses the cosmic conundrums of dark energy and the universe's accelerated expansion, suggesting a spacetime that is not static but evolves, where gravity's strength grows and diminishes, woven into the very fabric of the cosmos.

Theoretical Integration: Bridging Worlds

The integration of ECGT with quantum mechanics represents a frontier, promising yet fraught with complexities. This section delves into the intellectual challenges of marrying ECGT's macroscopic insights with the microscopic quantum world, exploring potential synergies and the theoretical leaps required to create a unified physical theory. These discussions are critical, highlighting both the potential and the hurdles in expanding our theoretical horizons.

Empirical Horizons: Evidence and Exploration

Within the empirical domains, ECGT stands poised to redefine our cosmic perceptions. This narrative recounts the theory's alignment with existing observational data and points to future experiments designed to peel back the layers of cosmic mystery. From probing the peripheries of black holes to unveiling the theoretical white holes, ECGT offers new lenses through which to view the universe.

A New Chapter in Cosmic Understanding

In the reflective calm of theoretical exploration, ECGT proposes a paradigm shift—a reimagining of gravity that challenges the foundational physics tenets. As we ponder the implications of ECGT, we call upon the collective curiosity and rigor of the scientific community. It is through this collaborative endeavor that we will either solidify ECGT's place in the annals of scientific advancement or refine our search for truth.

Section 1.3.5: Empirical Validation and Experimental Tests of ECGT

Unveiling the Canvas of Reality through Empirical Evidence

In the grand tapestry of theoretical physics, empirical evidence is the thread that weaves theory into the fabric of reality. For Evolving Compressive Gravity Theory (ECGT), rigorous empirical testing is the cornerstone of its journey from theoretical innovation to a cornerstone of cosmological understanding. This section outlines a blueprint for navigating this critical path, highlighting the experiments that could illuminate the predictions of ECGT and forge its credibility in the annals of science.

Glimpses of Support in Existing Observations

The universe has whispered hints of ECGT's veracity through various astronomical phenomena. Gravitational lensing anomalies, the enigmatic rotation curves of galaxies, and the uniformity of the cosmic microwave background radiation offer preliminary nods to ECGT's potential, suggesting a universe sculpted not merely by spacetime curvature but by dynamic, multidimensional gravitational interactions.

Crafting Experiments for a New Era of Gravitation

The quest to validate ECGT calls for experiments as innovative as the theory itself:

- **Defining Experimental Objectives: Our aim is to isolate and scrutinize the unique predictions of ECGT, such as the shielding effect of mass and the influences of an unseen fourth dimension on observable gravitational phenomena.**
- **Innovative Methodologies: Utilizing the latest in technology, from interferometry in gravitational wave observatories to precision measurements in particle physics, we stand on the precipice of uncovering the subtle effects predicted by ECGT.**

Proposed Key Experiments

These experiments are the crucibles in which the metal of ECGT will be tested:

- **Direct Tests of Gravitational Shielding: Tailored to detect variations in gravitational strength influenced by the shielding effect of mass, these experiments could reshape our understanding of gravitational interaction.**
- **Probing Extreme Gravitational Phenomena: The environments**

around black holes and the early universe serve as natural laboratories for observing the predictions of ECGT, offering a window into the most extreme conditions of the cosmos.

- **Bridging Gravity and Quantum Mechanics: Experiments at the quantum level may reveal how ECGT's predictions manifest on the smallest scales, potentially revolutionizing our understanding of quantum gravity.**

Navigating Challenges in the Laboratory of the Universe

Embarking on this empirical journey is not without its hurdles:

- **Technical Triumphs and Trials: The sophistication required in experimental design demands advancements in technology that can accurately measure the nuanced predictions of ECGT.**
- **Deciphering Cosmic Messages: Interpreting the data from these experiments will require a nuanced understanding, as we decipher subtle clues hidden in the noise of cosmic phenomena.**

Fostering Collaborative Frontiers

As we forge ahead, the path will be paved by the collaborative efforts of cosmologists, physicists, engineers, and technologists. These multidisciplinary teams are vital in designing and conducting the sophisticated experiments that will illuminate the truths of ECGT, blending expertise to ensure a holistic understanding of the results.

Reflections on the Horizon of Discovery

With each experimental endeavor, we inch closer to validating or refuting ECGT, potentially ushering in a new paradigm of gravitational physics. The successful validation of ECGT would not only redefine our understanding of the cosmos but also catalyze further scientific inquiries, inspiring a generation to look anew at the universe and its mysteries.

Section 1.3.6: Expanding Realms: The Transformative Simplicity of ECGT

Decoding the Cosmos with Elegant Simplicity

In the quest to unravel the mysteries of the universe, Evolving Compressive Gravity Theory (ECGT) emerges not merely as a beacon of innovation but as a harbinger of simplicity in our cosmic understanding. This theory reimagines gravity not as a fundamental force exerted by masses but as an emergent phenomenon arising from the nuanced interplay of mass and a hitherto unexplored fourth dimension. Here, we delve into the profound implications of this elegant theory, exploring how it promises to reshape our understanding of the universe's very fabric.

Redefining the Gravitational Dialogue

- **A Symphony of Forces: ECGT conducts a symphony where gravitational interactions are reinterpreted as the dance of differential shielding effects across a four-dimensional stage. This novel perspective not only harmonizes the disparate notes of quantum and cosmological scales but also simplifies the complex orchestration of fundamental forces.**
- **Universal Harmony: By proposing a unified mechanism for grav-**

itational interactions, ECGT offers a more streamlined narrative that aligns the celestial ballet of planets with the atomic waltz of quantum particles, underpinning both with a singular, elegant theory.

Cosmological Reawakening

- **Revising Cosmic Lore: With ECGT, the cosmos unveils its secrets in a new light, from the fiery forge of the Big Bang to the majestic swirl of galaxies. This theory offers a simpler explanation for the dark mysteries of matter and energy, casting aside the need for dark entities in favor of a dynamic, evolving gravitational force.**
- **Chronicles of the Cosmos: As ECGT rewrites the timeline of universal events, it invites us to reconsider the age and evolution of the cosmos, promising revisions that are more in tune with the observed universe than ever before.**

Quantum Continuity

- **Bridging Boundaries: ECGT elegantly bridges the seemingly insurmountable divide between the macroscopic laws of general relativity and the microscopic laws of quantum mechanics. It presents a unified vision that sees gravity as a consistent player across** all scales, offering a coherent framework that could one day culminate in the elusive theory of everything.

- **Quantum Rethink: The implications for quantum field theory are profound, as ECGT integrates gravitational phenomena into the quantum fold, potentially simplifying and unifying how forces interact at the most fundamental levels.**

Technological Renaissance

- **Innovative Horizons: The insights from ECGT are not confined to theoretical realms but extend their reach into the crucible of technology. Imagine propulsion systems that navigate the gradients of gravitational shielding or energy solutions that harness the dynamics of four-dimensional space—ECGT makes these sci-fi scenarios plausible.**
- **Engineering the Future: As we grasp the implications of ECGT, new avenues in aerospace design and materials science beckon, driven by a deeper understanding of how gravitational dynamics operate across dimensions.**

Philosophical Depths and Societal Echoes

- **Philosophical Paradigms: ECGT compels us to expand our philosophical horizons, challenging us to reimagine the essence of space and time and our place within this grand cosmic design.**

- **Societal Symphony: These conceptual shifts will resonate through society, influencing educational paradigms, cultural narratives, and perhaps even the existential questions we ponder about our universe and our role within it.**

The Call to Cosmic Curiosity

- **Future Frontiers: The path forward is illuminated by the promise of ECGT to revolutionize our understanding and exploration of the universe. It beckons the brightest minds to engage in a multi-disciplinary pursuit, testing, refining, and ultimately proving or disproving this bold new theory.**
- **A Unified Quest: As we stand on the precipice of potentially the greatest scientific revolution, ECGT not only offers answers but also ignites our collective imagination, urging us to dream of a universe where complexity yields to simplicity, and mystery gives way to understanding.**

Section 1.3.7: Quantum Gravity Echoes: Emerging Magnetic Mysteries of Calcium-48

Nuclear Insights and Gravitational Theory

Exploring the magnetic properties of calcium-48 provides unique insights into subatomic particle interactions, foundational to the Evolving Compressive Gravity Theory (ECGT). This approach sees gravity as a universally pervasive force, impacting all mass by influencing subatomic particles. This perspective aligns nuclear magnetic behavior with broader gravitational dynamics, deepening our understanding of gravity as a fundamental quantum phenomenon.

Bridging Nuclear Physics and Gravitational Theories

The study of calcium-48's magnetic properties extends our grasp of quantum gravity, essential to ECGT. This research suggests gravitational effects arise from collective subatomic behaviors, supporting the model where proximal masses are pushed towards their gravitosphere intersections, thus bridging nuclear phenomena with gravitational theories.

Enhancing Theoretical Depth and Robustness

Incorporating advanced nuclear physics insights, especially the magnetic properties of calcium-48, solidifies ECGT's foundation. This interdisciplinary approach not only reinforces the theory's robustness but also proposes a unified explanation of gravitational interactions across subatomic to cosmic scales, spurring novel hypotheses about the quantum underpinnings of gravitational forces.

Future Research and Implications

This integrated approach fosters new avenues for exploring how subatomic properties influence macroscopic gravitational dynamics. Future research, guided by ECGT's framework, could develop innovative methods to observe and manipulate gravitational interactions, deepening our understanding of the universe's fabric and potentially leading to groundbreaking discoveries in gravitational physics.

Chapter 1.4: Implications and Applications

Section 1.4.1: Implications for Galactic Structures and Dynamics

Reinterpreting Galactic Dynamics through ECGT

The rotation curves of galaxies and their distinctive structures often challenge traditional gravitational theories, which frequently resort to invoking dark matter to explain these phenomena. Evolving Compressive Gravity Theory (ECGT) reshapes this narrative, proposing that the intricate dance of galactic rotations and their structural dynamics are profoundly influenced by the rotational movements of central massive bodies, like black holes.

Unraveling Rotational Dynamics in Galactic Formations

At the heart of each galaxy, the rotational behavior of supermassive black holes acts not merely as a local phenomenon but as a cosmic architect of gravitational fields. These behemoths modulate gravitational pressures, with the most substantial effects manifesting at the equator where rotational speeds peak. This modulation results in diminished compressive gravitational pressure, promoting the formation of the discoid structures seen in galaxies. This elegant mechanism offers a natural explanation for galactic discs, eliminating the need for hypothetical dark matter.

Revisiting the Galactic Fountain Effect

The well-documented Galactic Fountain Effect, which describes the cyclical journey of gas between a galaxy's disk and halo, gains a new perspective under ECGT. The theory posits that gravitational pressures influenced by galactic rotation drive this cyclical gas motion, supporting dynamic stability and ongoing star formation within galaxies. This process underscores a natural recycling system within galaxies, propelled by the nuanced interplay of gravitational forces.

Linking Theoretical Insights to Observational Evidence

This discussion elucidates how ECGT revises our understanding of galactic dynamics and structure formation. It delineates the alignment of the theory's gravitational interaction predictions, influenced by rotational dynamics, with observed galactic formations and behaviors. By weaving together these insights, ECGT not only clarifies observed galactic phenomena but also offers predictions that are ripe for empirical testing.

Reflecting on Cosmic Dynamics

ECGT stretches beyond the confines of individual galaxies to suggest a transformative framework for comprehending the universe's structure and evolution. This theory challenges the scientific community to reconsider conventional cosmic models, armed with a robust set of theoretical tools that could revolutionize our understanding of the cosmos.

Section 1.4.2: Practical Implications of ECGT

Gravitational Dynamics Under ECGT

Overview of ECGT Principles

In the vast cosmos, where the mysteries of gravity interlace through space and time, ECGT introduces a groundbreaking perspective. By reconceptualizing gravity as a higher-dimensional compressive omnidirectional force that continuously permeates our universe and is locally absorbed by mass, ECGT transforms our understanding from a force that pulls to one that compacts. This absorption of gravitational force creates a lower-pressure region known as the gravitosphere around every mass, fundamentally altering how we perceive gravitational interactions.

Cosmic Dynamics and JWST Observations

Contemporary Insights from the Cosmos

The James Webb Space Telescope has ushered in a new era of astronomical discoveries, revealing distant galaxies with unprecedented clarity. These observations challenge traditional models that depend on dark matter to explain cosmic structures and dynamics. Instead, they provide fresh empirical support for alternative theories like ECGT, which offers explanations for these phenomena without the need for unseen matter.

Significance of Gravitosphere Intersections

At the core of ECGT is the concept of the gravitosphere, a spherical region surrounding a mass where compressive gravitational pressures are significantly reduced. When these gravitospheres intersect, the compressive force is further diminished, leading to the creation of unique gravitational hiatuses. These zones of reduced counterforce are pivotal as they fundamentally direct the paths celestial bodies navigate.

Detailed Mechanics of Gravitational Interactions

The Cosmic Arenas

Imagine gravitosphere intersections as cosmic arenas where the traditional rules of gravity are beautifully inverted. Here, the inherent compressive nature in space itself is subdued, reducing the gravitational push and clearing the stage for celestial movements. This phenomenon explains why cosmic entities are guided toward these intersections, like actors to a spotlight, seeking regions where the universe's embrace is significantly loosened.

A Journey Toward Serenity

Driven not by attraction but by the perpetual, omnipresent compressive force that permeates our universe, celestial bodies naturally gravitate toward zones where gravitational pressure is at its lowest. This movement mirrors the fundamental behavior of being pressed into spaces of minimal resistance, an elegant expansion of ECGT that narrates the grand story of

cosmic dynamics. This constant pressing force, experienced as weight on Earth, underscores the universal principle of matter seeking areas where it is least compressed.

Implications for Celestial Dynamics

Formation of Cosmic Structures

The principles laid out by ECGT illuminate the processes behind the formation of galaxies and other large-scale cosmic structures. The interactions at gravitosphere intersections can lead to the formation of distinctive structures such as spiral arms in galaxies or ringed formations in nebulae. These structures result from matter congregating along paths shaped by overlapping gravitospheres, influenced by the central mass's rotational momentum.

Star Formation and Stellar Dynamics

ECGT redefines the processes of star formation. As interstellar gases and dust are coaxed into these low-pressure zones by the universe's unseen hands, they gather and compress, eventually igniting to form new stars. The angular momentum of these nascent stars, and their subsequent planetary systems, is influenced by a combination of factors: the interaction of galactic centrifugal forces, directional flows from expanding equatorial gravitosphere intersections, and the compacting effects of their own gravitospheres.

Linking Theory to Observable Phenomena

Predictive Power of ECGT

Armed with ECGT, astronomers peer into the cosmos with renewed vision, predicting where stars might form or galaxies twist, guided by maps of gravitosphere intersections. This theory does not merely suggest where to look; it illuminates the path, offering a predictive compass by which to navigate the celestial unknown.

Observational Echoes of Theory

The data streaming in from the void—light bent in improbable arcs, stars born in unexpected locales—speak to the accuracy of ECGT. These observational echoes affirm the theory's predictions, suggesting that we are only beginning to uncover the depth of the universe's secrets.

Section 1.4.3: Revising Astrophysical Theories

Cosmic Puzzles and Traditional Models

In the grand cosmic theater, traditional astrophysical models have long served as our guides to the stars, galaxies, and the enigmatic realms of black holes and neutron stars. These models, foundational yet filled with unanswered questions about dark matter, dark energy, and the origins of gravitational waves, prompt us to expand our understanding beyond the conventional.

Stellar and Galactic Dynamics: A New Vision with ECGT

Transforming Our View of Star Formation

Imagine a universe where the mysterious flat rotation curves of galaxies no longer hint at invisible dark matter but reveal the elegant dynamics of ECGT. Here, gravitational dynamics are not governed by unseen forces but are shaped by the differential absorption of a universal compressive force, reshaping our understanding from the ground up.

The Luminous Dance of Galaxies

Within the framework of ECGT, the Tully-Fisher relation—a cosmic dance linking a galaxy's luminosity to its swirling arms—finds a new rhythm. The diameter of a galaxy's gravitospheric disk, sculpted by the central mass's angular momentum, nurtures the birth of stars whose light bathes the cosmos. This relationship, once attributed to dark matter, now sings of the interactions within gravitospheres, illuminating our night sky with newfound clarity.

The Enigma of Black Holes and Neutron Stars

Redefining Horizons and Singularities

Journey deeper into the heart of darkness—black holes under ECGT are no longer realms of singularity and loss but areas where the sensation of gravity softens. What we once viewed as the final curtain, the event horizon, becomes a dynamic stage where gravitational forces are modulated, offering a fresh perspective on these cosmic enigmas.

A New Tale of the Information Paradox

Far from the irreversible oblivion of information, black holes as envisioned by ECGT become archives of cosmic history. Here, information is not destroyed but intricately transformed and redistributed across the tapestry of intersecting gravitospheres, challenging the very notion of information loss in the stark depths of black holes and neutron stars.

The Whisper of Gravitational Waves

Unveiling New Sources

In the cosmos shaped by ECGT, gravitational waves whisper tales of new beginnings, not just from the cataclysmic dances of binary stars but from the serene meetings of intersecting gravitospheres. These novel sources, revealed through the lens of ECGT, expand our understanding and detection of the universe's rhythmic pulsations.

Future Horizons for Observatories

As the predictions of ECGT stretch the canvas of known physics, observatories like LIGO must tune their senses finer, adapting to capture the subtle dialogues of gravitational waves newly understood. This adaptation promises to not only validate ECGT but also to expand the realms of what can be observed in the cosmic void.

Epilogue: Toward a New Understanding

As we delve through the implications of ECGT, it becomes evident that this theory does not merely adjust our cosmic perceptions but revolutionizes them. It invites us to reconsider our universe, offering new frameworks for understanding the dance of galaxies, the secrets of black holes, and the symphony of gravitational waves. ECGT promises a profound shift in our cosmic narrative, urging us to look anew at the old mysteries that have long captivated our questing minds.

Section 1.4.4: Addressing Dark Matter, Dark Energy, and Quantum Integration

Alternative Explanations for Dark Matter

In the cosmic dance of galaxies, where mysterious forces seem to tug at the fabric of the universe, traditional astrophysics posits dark matter as an invisible player. Yet, ECGT invites us to view these phenomena through a new lens. It suggests that the enigmatic flat rotation curves of galaxies and the peculiar bending of light—gravitational lensing—are not the works of unseen matter but the result of a universe orchestrated by differential gravitational forces, where each celestial body contributes to the cosmic melody through its own gravitosphere.

Galaxy Rotation Curves:

Imagine galaxies not as entities held in spin by invisible matter, but as dynamic systems where gravitospheres interact to create the observable rotation curves. ECGT reinterprets these curves as a manifestation of gravitational forces modulated across galactic expanses, challenging the dark matter paradigm with a symphony of observable, dynamic interactions.

Gravitational Lensing:

Shedding the need for dark matter, ECGT explains the bending of light around massive objects as a dance of intersecting gravitospheres. Here, photons travel through a cosmos where their paths are sculpted not by invisible mass but by the palpable pressures of gravitational fields they encounter, redefining our understanding of cosmic visibility.

ECGT and Cosmic Structure Formation

A New Framework for Understanding the Universe

ECGT reinterprets the fabric of the cosmos by proposing dynamic gravitational interactions shaped by a variable gravitational constant, $G(t)$**. This innovative approach explains the formation and behavior of galaxies as observed by JWST, demonstrating that gravitational forces, influenced by the properties of four-dimensional space, can account for the complex structures and motions observed in the universe without resorting to dark matter.**

Rethinking Dark Energy

Dark energy, the mysterious force propelling the universe's accelerating expansion, is traditionally viewed as a uniform energy suffusing space. ECGT, however, paints a different picture—suggesting that what we perceive as dark energy is the result of an evolving gravitational force, a natural outcome of a universe in perpetual motion.

Accelerating Expansion of the Universe:

Under ECGT, the accelerating expansion is seen not as a push by an unknown energy but as an inevitable consequence of a universe whose gravitational dynamics are continuously evolving, reshaping our cosmological narrative into one of dynamic change rather than static presence.

Quantum Gravity ECGT Style

Delving into the quantum realm, ECGT extends its reach to the microscopic interactions of particles, suggesting a universe where every particle, no matter how small, is influenced by gravitational interactions.

Impact on Quantum Mechanics:

ECGT introduces a scenario where gravitospheres at the quantum level influence particle behavior, providing a gravitational underpinning to phenomena typically governed by quantum mechanics. This integration hints at a universe where the rules of gravity are not sidelined but are integral even at the smallest scales, offering new insights into particle interactions and the fundamental forces that dictate their existence.

Specific Explanation of Allowed States and Molecular Configurations:

In this enriched quantum landscape, ECGT proposes that the unique shapes and interactions of gravitospheres dictate the allowed states of particles and the configurations of molecules, suggesting a gravitational component to chemical bonds and molecular structures, thus broadening the horizons of chemistry and materials science

Epilogue: A New Cosmic Perspective

In the tapestry of this discourse, we've unraveled the mysteries shrouding dark matter and dark energy, revealing a universe illuminated by the coherent principles of ECGT. Here, the celestial phenomena once attributed to invisible forces are reinterpreted as the dynamic interplay of gravitational fields, visible through the lens of evolving gravitospheres. As ECGT peels back the cosmic veil, we are invited into a realm where every particle, every star, and every galaxy is interconnected in a dance dictated by the very forces we are now beginning to understand. This journey through ECGT not only challenges our longstanding paradigms but also enriches our cosmic narrative, promising a future where the mysteries of the universe are not locked away in the shadows of the unknown but are accessible, compelling, and deeply integrated into our understanding of the natural world.

Section 1.4.5: Speculative Applications in Understanding the Universe

New Methods for Cosmological Observation

In the expansive arena of cosmic exploration, ECGT presents theoretical insights that could radically transform our methods of observing and measuring the universe. By suggesting the presence of more subtle gravitational interactions than currently acknowledged, ECGT paves the way for developing new observational techniques that could unlock a deeper understanding of cosmological phenomena.

- **Gravitospheric Mapping: One innovative approach inspired by ECGT is the mapping of gravitospheres, which could reveal intricate gravitational interactions between celestial bodies. This method would allow astronomers to visualize and study the dynamic gravitational forces at play across various scales, from planetary systems to entire galaxies.**
- **Detection of Gravitational Variations: ECGT also motivates the development of technologies to detect and measure subtle variations in gravitational fields. These measurements could provide crucial data for testing the predictions of ECGT and enhance our understanding of cosmic structure and behavior.**

Technological Innovations

The implications of ECGT extend beyond traditional observational methods, inspiring technological innovations that leverage quantum mechanics and the properties of gravitational fields to pioneer advanced sensing capabilities.

- **Advancements in Sensor Technology: The integration of quantum technology, particularly through the use of quantum accelerometers, marks a significant advancement in sensor capabilities. These devices, which utilize Bose-Einstein condensates, offer sensitivity to acceleration that far exceeds that of traditional sensors.**
- **Quantum Navigation Trial: A notable trial in the United Kingdom tested quantum navigational technology on an aircraft, showcasing the potential of these sensors to operate independently of satellite-based systems like GPS. This trial used delicate quantum technologies to measure precise movements without external references, suggesting a future where navigation could rely entirely on intrinsic gravitational sensors.**
- **Quantum Accelerometers: Employing Bose-Einstein condensates, these accelerometers can detect minute changes in movement by manipulating a supercooled cloud of rubidium atoms. This technology not only has implications for navigation but could also prove essential in measuring the effects of the fourth-dimensional space proposed by ECGT.**

- **Potential Applications: Initially aimed at enhancing autonomous shipping, the broader applications for quantum accelerometers include autonomous mining, industrial settings, and eventually personal autonomous vehicles. These devices could become fundamental tools for industries where traditional GPS systems are inadequate or unreliable.**

Quantum Gravity Untangled

ECGT proposes a model where even the smallest quantum particles are influenced by gravitational interactions, suggesting a world where quantum mechanics and gravitational theories are not separate but interconnected.

- **Impact on Quantum Mechanics: By incorporating gravitational interactions into the quantum realm, ECGT could lead to a profound revision of how we understand particle behavior and inter-particle forces at the smallest scales.**
- **Specific Explanation of Allowed States and Molecular Configurations: ECGT could provide new insights into why particles and molecules assume specific configurations, potentially revolutionizing fields from chemistry to materials science.**

Bridging Microscopic and Macroscopic Physics

ECGT stands as a bridge between the cosmic scales of general relativity and the minute scales of quantum mechanics, offering a unified understanding that spans the breadth of physical phenomena.

Epilogue: Pioneering the Future

So now here we have revealed the profound potential of ECGT to not only enhance our theoretical understanding but also drive significant technological and observational innovations. By pushing the boundaries of what is possible, ECGT encourages us to rethink our place in the universe and our approach to unraveling its mysteries.

Section 1.4.6: Interdisciplinary Impacts

The Cosmic Ripple of ECGT

As ECGT unfurls its theoretical tendrils across the fabric of science, its influence stretches beyond the realms of physics, reaching into the intricate interplay of astrophysics, astrobiology, and beyond.

- **Astrophysics and Astrobiology: Picture a universe where ECGT rewrites the laws governing the celestial ballet of planets and stars. This new doctrine explains not just the mechanics of galaxies but paints a broader canvas illustrating the potential habitats for life across the cosmos. It challenges us to rethink our search for extraterrestrial life, considering diverse gravitational environments as nurseries of potential life.**
- **Particle Physics and Quantum Mechanics: Delve deeper into the quantum forest where ECGT sheds light on the shadowy interactions at the subatomic level. Here, gravitational forces whisper among the particles, suggesting a profound connection between the cosmic dance and the quantum leap.**
- **Computational Sciences and Engineering: ECGT inspires a revolution in computational models and engineering practices. Envision more accurate simulations that predict celestial dynamics or structural integrity under diverse gravitational scenarios, enhancing our ability to design, forecast, and innovate.**

Philosophical Echoes and Educational Waves

ECGT not only revises our scientific understanding but resonates through the corridors of philosophy and education, urging a reevaluation of our very framework of understanding.

- **Philosophical Echoes: ECGT challenges our existential musings on space and time, urging philosophers to weave new narratives around the nature of reality. It invites us to ponder a universe where space-time is not a passive arena but an active participant in the cosmos.**
- **Educational Waves: As ECGT percolates through educational curricula, it enriches scientific literacy, equipping the next generation of thinkers with a more nuanced understanding of the universe. It proposes a curriculum where gravitational theories are not mere footnotes but central chapters in the story of science.**

Expanding Scientific Horizons

ECGT's influence burgeons into chemistry and biology, providing new frameworks for understanding the molecular and evolutionary tapestries of life.

- **Chemistry and Biology: Imagine chemistry where bonds are not just electrical but gravitational, and biology where evolutionary trajectories are guided by the subtle forces of a dynamic universe.**
- **Societal Transformations: As we align with ECGT principles, we stand on the cusp of societal transformation. This theory does not just predict physical phenomena but forecasts a shift in human evolution, heralding the age of Homo Gnosis—a new epoch where our enhanced understanding and capabilities redefine our relationship with the universe and each other.**

Epilogue: A Unified Scientific Perspective

In the vast expanse of knowledge, ECGT emerges as a beacon, guiding us to a more integrated understanding of the universe. It calls upon us to embrace a holistic view of science, where disciplines converge and collaborate, fostering a future rich with discovery and enlightenment.

Section 1.4.7: Future Directions for Research and Application

Charting the Course Ahead

As we stand on the brink of new frontiers in gravitational research, ECGT offers a roadmap that promises to illuminate uncharted territories of understanding and application. This roadmap is not just a guide but a call to adventure—a challenge to the scientific community to pioneer the depths and breadths of a universe more intricately woven than previously imagined.

- **Identifying Key Research Areas: The quest begins with defining the domains where ECGT will have the most profound impact. Theoretical advancements need to weave together the very fabric of our understanding, tying loose ends and strengthening the bonds between observed phenomena and theoretical predictions. Empirical testing will serve as our compass, guiding us through the validation of these theories with precise measurements and careful observations, charting stars and the spaces between them.**
- **Technological Developments: Our journey is supported by the tools we forge along the way—innovative technologies that not only detect the subtle whispers of gravitational changes but also harness these forces in ways that transcend current capabilities.**

These tools will open new doors to applications that once resided firmly in the realm of science fiction.

- The Graviton Detection Experiment: In the enigmatic theater of gravitational research, a bold new experimental saga is set to unfold, potentially illuminating or dispelling the shadows cast by the elusive gravitons—those theoretical sprites of quantum gravity lore. Picture a massive aluminum bar, a dormant leviathan weighing nearly 4,000 pounds, chilling on the brink of absolute zero. This frosty giant, cradled by the most avant-garde yet unborn quantum sensors, awaits the faintest whispers of the universe—vibrations that could herald the presence of gravitons. Invoking the mystical 'gravito-phononic effect', akin to the legendary photoelectric effect, this bar hopes to translate the cosmic dance of neutron star collisions into discrete quantum murals. As these sensors stand guard, decoding the subtle murmurs of gravitational waves, we teeter on the edge of potentially monumental discoveries. Will we catch a glimpse of these phantom particles, or will we bolster ECGT's audacious claim of their nonexistence? Each scenario weaves its own narrative thread, promising to either unravel or embroider the rich tapestry of ECGT, driving us ever deeper into the mysteries of quantum gravity and setting the stage for a new era of theoretical and technological marvels.

Building Bridges Across Disciplines

The path to understanding ECGT's full potential is paved with collaboration—interdisciplinary, international, and interstellar.

- **Global Cooperation: In the spirit of unity, global partnerships will emerge as cornerstones of our exploration. By pooling resources, sharing knowledge, and synchronizing our efforts, we ensure that the insights and advancements under ECGT are not confined by borders but are enriched by diverse perspectives.**
- **Funding and Resources: To voyage far, the engines of research and development must be fueled adequately. Here lies the role of visionary funding bodies—partners in our quest, who see the value in investing in a future shaped by deeper understanding and broader applications of gravitational theory.**

Envisioning Practical Applications

The implications of ECGT ripple through various sectors, promising to revolutionize industries and elevate our technological prowess to new heights.

- **Aerospace and Beyond: Imagine spacecraft navigating by the very fabric of space itself, propelled by the manipulation of gravitational fields. ECGT could turn the vacuum of space into a canvas of** opportunities—each star, each planet, a node in the vast network of cosmic travel.

- **Environmental Stewardship: On Earth, ECGT offers new methods to predict and perhaps even mitigate the caprices of nature. By understanding and maybe influencing the gravitational nuances of our planet, we could better prepare for and respond to environmental challenges.**

Epilogue: A New Era of Discovery

In this grand tapestry of exploration, each thread of ECGT is intertwined with countless possibilities. We are not just observers of the universe but active participants in its unfolding story. As we embrace ECGT, we are rewriting the cosmic playbook—expanding our capabilities, deepening our understanding, and potentially altering the course of human evolution.

- **Call to Action: Let us then move forward with purpose and passion. The journey of ECGT is not for the faint of heart, but for the bold, the curious, and the visionary. Together, we can turn the mysteries of today into the breakthroughs of tomorrow, and in doing so, reshape our world and beyond.**

Section 1.4.8: Epilogue: Charting New Horizons

Weaving the Cosmic Fabric

As our exploration of ECGT approaches its crescendo, we reflect on the profound shifts it heralds—a paradigm where the cosmos is not simply a space filled with gravitational attractions but a dynamic tapestry woven with the threads of four-dimensional interactions.

Impact on Theoretical Physics

ECGT has profoundly challenged the foundational aspects of gravitational theory, proposing a radical new framework. The force of gravity is no longer understood as a pulling attraction emanating from a mass; rather, it is envisioned as a gradually evolving, continuous, pervasive, omnidirectional compressive force—a subtle yet powerful pressure that permeates our universe. This pressure, arising from the intricate ballet of dynamic acceleration within four-dimensional space, is uniquely modified by the presence of each mass. This paradigm, which embraces us and the cosmos, promises to unify disparate phenomena under a single theoretical umbrella, offering more cohesive and comprehensive explanations for the behaviors observed across the vastness of space.

Empirical Support and Future Directions

Linking Observations with Theoretical Predictions

The data from JWST validate key aspects of ECGT, showing that the unexpected behaviors of galaxies and their rotation curves can be explained by changes in gravitational dynamics proposed by the theory. This alignment between observation and theory not only strengthens the case for ECGT but also sets the stage for future astronomical research. As we continue to observe the universe with JWST and other telescopes, we expect more evidence to emerge that will further corroborate ECGT's predictions, potentially leading to its broader acceptance and integration into the standard cosmological model.

A Clarion Call to the Scientific Community

In its narrative of cosmic evolution, ECGT stands as a beacon calling for a collective voyage of discovery and innovation. We summon the global scientific community to embark on this journey, to engage in robust collaboration and rigorous exploration that transcend traditional boundaries.

Fostering Collaboration and Innovation

We advocate for collaborative exploration that spans disciplines, cultures, and continents. The pursuit of understanding ECGT's full potential is not just a scientific endeavor but a global imperative that demands the pooling of resources, sharing of knowledge, and synchronization of efforts across the globe. It is through such united endeavors that the mysteries of ECGT can be fully unraveled, paving the way for groundbreaking discoveries that could redefine our understanding of the universe and fundamentally reshape our lives.

Envisioning the Future

As we chart new horizons, ECGT beckons us to envision a future replete with possibilities—where our understanding of gravity not only enhances our knowledge of the universe but also revolutionizes technologies, reimagines educational curricula, and redefines our philosophical outlook.

Inspiring the Next Generation

Let this be our legacy—to instill in future generations a curiosity that is boundless and a determination that is unwavering. We are on the brink of a new era in scientific inquiry, one that will require the brightest minds to continue the work begun here. By embedding the principles of ECGT into the very fabric of our educational systems, we prepare our successors to wield the tools of tomorrow, ensuring that the journey of discovery is ever onward.

Cosmic Closure

As this chapter closes, the dialogue about ECGT is just beginning. Its potential to reshape our cosmic understanding is immense. With hearts courageous and minds open, let us stride forward into this uncharted territory, ready to uncover the secrets that lie waiting within its cosmic embrace.

Chapter 1.5: Revising the Age of the Universe

Section 1.5.1: Unveiling New Beginnings

Enigmatic Shadows in Cosmic Theories

The Puzzle of Dark Matter and Dark Energy: In the grand cosmic theater, dark matter and dark energy have long played the roles of invisible architects, molding the universe's structure and fate with unseen hands. Introduced to reconcile observed astronomical phenomena with gravitational theories, these entities remain elusive, hidden from direct observation. The recent revelations by the James Webb Space Telescope (JWST) further complicate this narrative, capturing distant galaxies that challenge our understanding of the universe's age and expansion.

Revealing Cracks in the Cosmic Canvas

Challenging Conventional Wisdom: The reliance on dark matter and dark energy has woven a cosmological model that, though useful, is frayed with inconsistencies. The cosmological constant problem highlights severe discrepancies within dark energy discussions, and the JWST's observations of galaxies like JADES-GS-z14-0—far beyond the expected

cosmological horizon—suggest our models of cosmic expansion may need reevaluation, and even lead to speculation that events such as the Cosmic Microwave Background (CMB) may have occurred at different times in different locations of our universe. These findings underscore the limitations of current theories in explaining the accelerating expansion and distribution of mass-energy across the cosmos.

A New Perspective in the Cosmos

Presenting Evolving Compressive Gravity Theory (ECGT): As we reassess traditional gravitational theories, ECGT emerges as a bold new chapter. Proposing a dynamic gravitational constant, G(t) , that evolves with the universe's rhythm, ECGT offers a compelling alternative. This theory redefines gravitational interactions as dynamic interplays between a four-dimensional force and uniquely shaped gravitospheres surrounding each mass. Where these gravitospheres overlap, masses are driven not by attraction but by a push toward regions of shielded pressure, providing a clear, empirical framework that aligns with the JWST's groundbreaking observations.

Empirical Alignments

ECGT and the Empirical Universe: ECGT's premise that G(t) can vary aligns naturally with the JWST findings, offering a robust explanation for the observed phenomena without relying on dark matter or dark energy. It explains the extended visibility of distant galaxies through its model of gravitational interaction, suggesting that the universe's expansion allows us to see light from greater distances than previously thought possible, due to periods of rapid expansion influenced by dynamic gravitational changes.

Redefining the Forces that Bind the Cosmos

A Paradigm Shift in Gravitational Understanding: With ECGT, gravity sheds its cloak of mystery, revealing itself not as a force of pulling attraction but as a constant pervasive, compressive pressure. This viewpoint shift from pulling forces to an omnidirectional push towards the shielded places near masses invites us to rethink the fundamental processes shaping the cosmos. It opens new pathways for solving long-standing puzzles and paves the way for future explorations into the universe's true nature and origins.

Section 1.5.2: The Cosmic Tapestry: Weaving Data with Theory

Decoding the Cosmic Microwave Background (CMB)

The Conventional Canvas and Its Incomplete Picture

Traditionally, the Cosmic Microwave Background (CMB) is celebrated as the luminous echo of the Big Bang—a pristine snapshot captured a mere 380,000 years post-creation. This radiant relic has served as a cornerstone in validating the classical Big Bang model, casting a uniform glow across our cosmic understanding. Yet, beneath this surface of apparent simplicity, lie profound complexities that our conventional models struggle to illuminate fully.

Challenges to Uniformity

Recent telescopic triumphs, particularly the unprecedented observations by the James Webb Space Telescope (JWST), have begun to unravel the uniform narrative traditionally woven around the CMB. These observations challenge the traditional 13.8 billion-year horizon, suggesting complexities in cosmological distance measurements that may imply a more nuanced understanding of the universe's expansion. This emerging perspective, nurtured by Evolving Compressive Gravity Theory (ECGT), suggests that the gravitational forces during the CMB epoch might have been significantly less intense, reshaping our understanding of cosmic structure formation in profound ways.

Providing These Possibilities

Reimagining Gravity

ECGT redefines gravity as a dynamic interplay between a pervasive, omnidirectional compressive force originating from a higher dimensional reality—akin to a form of acceleration—and the regions of reduced pressure near masses where this force is absorbed. This model, eschewing the traditional notion of gravitational pull, posits that gravitational effects arise not from attraction but from the decreased counter pressure experienced at the intersections of these absorption regions, allowing the force of compressive gravity to push masses toward each other.

Integrating $G(t)$ into Cosmological Calculations

Revolutionizing Cosmic Calculations

It is precisely because ECGT reassigns the source of gravity away from being an integral primary function of mass that it becomes possible to introduce a variable gravitational constant, $G(t)$, into cosmological calculations, marking a pivotal shift in our scientific saga. This innovative approach redefines our understanding of gravitational forces, influencing everything from the birth of galaxies to the expansive behavior of the cosmos. By recalibrating the traditional models with $G(t)$, ECGT offers a lens that more accurately reflects the observed intricacies of cosmic expansion and structure formation.

Eliminating Dark Matter from Cosmological Calculations

Instead of relying on dark matter as a theoretical filler to balance gravitational equations, ECGT recalibrates the gravitational landscape by viewing gravity as a variable force. This shift allows us to recompute cosmic parameters such as the age of the universe more directly, providing a clearer and more accurate understanding based on observable data, rather than theoretical constructs.

Future Directions and the Role of Emerging Data

Empirical Horizons

The path forward in validating ECGT lies in the empirical realm—through meticulous observations of the CMB and other cosmic phenomena. Advanced technologies and international collaboration will be crucial in gathering the precise data required to test the predictions of ECGT, turning theory into observable fact.

Charting New Cosmic Courses

As we refine our cosmic comprehension through the lens of ECGT, we edge closer to a unified theory that not only explains but also predicts the phenomena of the universe with unprecedented accuracy. The integration of $G(t)$ into cosmological models is not merely theoretical—it is a necessary evolution that promises to illuminate the darkest corners of our universe.

Section 1.5.3: Reassessing Cosmic Milestones

A New Gaze into the Cosmic Kaleidoscope

Challenging Traditional Cosmological Canvas

In the grand cosmic theater, where the luminescence of galaxies unfolds ancient narratives, traditional models wielding the constants of dark matter and dark energy have sculpted our understanding of the universe's expanse. These models, anchored in a static gravitational constant, now face the piercing scrutiny of Evolving Compressive Gravity Theory (ECGT), which establishes a new narrative across the cosmos by proposing a dynamic $G(t)$, evolving gracefully with the universe's rhythm.

Harmonizing Cosmic Dissonance

The Symphony of Gravitational Waves

From the clandestine dance of matter and antimatter, where particles meet their mirrored fates, springs a symphony of gravitational waves, charting a new frontier for cosmic exploration. ECGT posits these annihilation events as vital sources of gravitational waves, offering a unique sonata of empirical evidence to test the theory's bold claims. Laboratories across the globe, buzzing with the energy of discovery, replicate these cosmic interactions, employing the latest sophisticated quantum technologies, such as Bose-Einstein Condensates, that measure the subtlest whispers of gravitational change. These findings could illuminate the path

laid out by ECGT, affirming or refining its narrative about the nature of gravitational forces.

Laboratory Frontiers: Crafting Cosmic Events

Innovative experiments now seek to mimic the universe's most dramatic ballets—matter-antimatter annihilations—within the confines of earthly labs. By observing the resulting gravitational perturbations, scientists aim to capture the echoes of these events, providing insights into the gravitational nuances unleashed by such profound encounters.

Revising the Cosmic Codex

Recalculating Cosmic Chronicles

With the evolving gravitational constant $G(t)$ as our guide, the storied past of the universe undergoes a meticulous revision. This recalibration, a testament to ECGT's innovative spirit, offers a more dynamic portrayal of cosmic evolution, suggesting that the milestones of cosmic history—from the formation of galaxies to the expansive ballet of the cosmos—may have played out under different gravitational tempos than previously scored. Subject to a different master conducting gravitational interactions, these cosmic events are further influenced by the unique gravitosphere shapes that arise from the translational and angular momentum of their associated masses, adding new layers of complexity and predictability to our understanding of gravitational dynamics.

Empirical Concerto: A Call for Global Harmony

The future of ECGT hinges on a global symphony of collaborative research. International scientific ensembles, from gravitational wave hunters to cosmic microwave background analysts, are poised to test, challenge, and potentially validate the transformative perspectives offered by ECGT.

Philosophical Renaissance in Cosmic Understanding

Rethinking the Universe's Overture

As we stand at the precipice of cosmic reawakening, ECGT beckons us to redefine the narratives of celestial evolution. This theory not only revises our scientific texts but also reinvigorates the educational and philosophical underpinnings of cosmology, urging a new generation of thinkers to embrace the mysteries of the universe with open arms and curious minds.

Epilogue: Charting the Uncharted

In the quest to understand our universe's deepest secrets, ECGT emerges as a beacon of innovation, challenging us to gaze beyond the visible and measurable—to explore the subtle and profound forces that sculpt the cosmos. As this chapter of cosmic exploration unfolds, we are reminded that every question answered leads us to new mysteries, and every mystery resolved opens new realms of inquiry.

Section 1.5.4: Blueprint for Tomorrow: Charting Future Research

Navigating the Unknown

Phased Research Plan:

- **Phase 1: Theoretical Consolidation - Dive deep into refining the mathematical bedrock of ECGT, wielding computational models to simulate a universe where gravitational strengths shift.**
- **Phase 2: Empirical Testing - Craft and execute key experiments designed to catch the elusive gravitational effects ECGT predicts, from novel gravitational wave sources to unique gravitational lensing phenomena.**
- **Phase 3: Data Integration and Analysis - Amass and dissect data from across the globe, fine-tuning the theoretical constructs to ensure they resonate with the empirical truths of the cosmos.**

Key Experiments:

- **Laboratory-Based Experiments - Harness high-energy particle accelerators to stage and study matter-antimatter interactions, watching closely for their gravitational tales.**

- **Astronomical Observations - Join hands with international observatories, seeking out cosmic phenomena that whisper the secrets of variable gravitational strengths.**
- **Interdisciplinary Approaches - Marry insights from quantum physics and cosmology to delve into how changing gravitational constants might sculpt the universe at its most fundamental levels.**

Collaborative Frontiers

Global Collaboration:

- **Establishing International Consortia - Forge alliances that span continents and disciplines, pooling resources and harmonizing methods in a shared quest for knowledge.**
- **Shared Data Repositories - Construct digital commons where data flows freely, allowing researchers from every corner of the world to contribute to and benefit from a collective pool of knowledge.**
- **Technology Development - Collaborate on forging new sensors and instruments, finely tuned to detect the subtle shifts in gravitational forces that ECGT anticipates.**

Expertise and Resource Allocation:

- **Interdisciplinary Teams - Blend expertise from astrophysics, quantum mechanics, and computational modeling to confront the intricate puzzles of ECGT.**

- **Funding Strategies - Cultivate financial support from global foundations and governmental bodies, fueling a long-term quest to decipher the universe's gravitationally bound mysteries.**
- **Educational Programs - Design and deliver specialized training and workshops, enlisting the Cosmic Chronicles Tetralogy and the Cosmic Chronicles Navigator - GPT as primary tools to ignite curiosity and forge understanding across generations.**

Technological Innovations:

- **Advances in Measurement Technologies - Push the boundaries of what's measurable with pioneering quantum sensors that can trace the faintest gravitational fluctuations, like those recently crafted by international research teams. These tools stand ready to chart new territories of quantum and gravitational interactions, providing crucial support for ECGT's bold predictions.**
- **Simulation Software - Invest in cutting-edge simulation software designed to visualize cosmic events as they might occur under varying gravitational conditions, offering a digital stage where theoretical predictions meet empirical tests.**

Section 1.5.5: Cosmic Synergies: Observations and Predictions

Empirical Alignments

Upcoming Astronomical Missions

- **Charting New Frontiers: Future space missions, equipped with next-generation telescopes and interstellar probes, are poised to extend our cosmic map. These missions can be specifically aligned to explore the gravitational phenomena predicted by ECGT, potentially transforming abstract theoretical models into observable realities.**
- **Deciphering Cosmic Structures: The data from celestial giants like galaxy clusters, black holes, and neutron stars hold keys to unlocking the secrets of the cosmically evolving gravitational constant, G(t). Observations of these cosmic entities could shine a light on the dynamic nature of gravity as envisioned by ECGT, challenging the static views held by conventional theories.**

- **Quantum Gravity Laboratories: Around the globe, laboratories are pushing the boundaries of how we measure gravitational forces at the quantum scale. These facilities could soon provide empirical evidence to support or challenge ECGT's predictions about the interactions of gravitospheres and their influence on gravitational dynamics.**
- **Global Telescopic Collaborations: International observatories are crucial in gathering the comprehensive data needed to test the far-reaching implications of ECGT. By collaboratively analyzing gravitational lensing and wave detections, scientists can forge a new understanding of cosmic expansion and structure formation.**
- **Development of Gravitational Telescopes: Pioneering the frontier of gravitational observation, the proposal for gravitational telescopes is set to redefine our cosmic explorations. These specialized instruments are designed to delve into the gravitational landscape of the universe by mapping the distinct sizes and shapes of gravitospheres associated with various cosmic masses. By providing detailed visualizations of these gravitational fields, gravitational telescopes aim to offer unprecedented insights and empirical evidence that could validate or refine the predictions made by Evolving Compressive Gravity Theory (ECGT). The deployment of these telescopes would not only enhance our understanding of gravitational dynamics but also enable direct testing of ECGT on**

an astronomical scale, potentially revolutionizing our understanding of cosmic structure and interaction.

Gravitospheric Shape Effects

- **From Cosmic to Quantum: The unique shapes of gravitospheres, influenced by the translational and angular momentum of their associated masses, offer a novel perspective on gravitational interactions. Understanding these shapes could revolutionize our interpretations of gravitational effects, from the largest celestial bodies to the tiniest quantum particles.**
- **Simulations of Cosmic Dances: Advanced computational models now simulate how these gravitospheres interact, predicting phenomena that bridge the gap between theory and observation, thereby providing a crucial testbed for ECGT.**

From Theory to Practice and Back

Technological Innovations

- **Revolutionary Sensing Technologies: The development of new sensors capable of detecting the faintest gravitational changes promises to bring the subtle predictions of ECGT into the empirical realm. These technologies, emerging from the fusion of quantum mechanics and gravitational research, could dramatically enhance our ability to observe the universe.**

- **Enhanced Observational Tools: Innovations in telescope technology and observational methodologies are not just advancements; they are necessities for probing the depths of space as dictated by ECGT. These tools will be instrumental in validating the theory's predictions about gravitational interactions across the cosmos.**

Strengthening ECGT

- **Empirical-Theoretical Feedback Loop: As new data streams in from these advanced technologies and collaborative efforts, they feed back into the theoretical framework of ECGT. Each empirical finding refines the theory, sharpening its predictions and broadening its applicability.**
- **Real-World Applications: Beyond the lofty realms of theoretical physics, the practical applications of a confirmed ECGT are boundless. From revolutionizing navigation systems that rely on gravitational measurements to rethinking how we communicate across vast distances in space, the implications are as practical as they are profound.**

Fostering Global Scientific Dialogue

- **Cultivating Collaborative Discoveries: The path to fully understanding ECGT is paved with global collaboration. Regular international conferences, symposia, and workshops dedicated to gravitational research will be vital in fostering a fertile environment for discussion, debate, and discovery.**
- **Educational Integration: By weaving the principles of ECGT into educational curricula through tools like the Cosmic Chronicles Tetralogy and the Cosmic Chronicles Navigator - GPT, we prepare the next generation of scientists to continue this exploratory journey, ensuring the legacy of ECGT extends far into the future.**

Section 1.5.6: Reflections on Infinity: Philosophical and Educational Implications

Philosophical Paradigms

- **Expanding Universe, Expanding Minds: ECGT opens up new vistas in our cosmic understanding, challenging us to stretch our perceptions alongside the ever-expanding universe. This theory redefines gravity not merely as a force but as a dynamic, evolving presence, urging us to reconsider our relationship with the cosmos. It paints a picture of a universe in perpetual creation, molded by the interplay of mass and energy, sparking profound reflections on our interconnected existence within this grand tapestry.**
- **Rethinking Our Cosmic Place: ECGT reshapes foundational notions of the universe, influencing deep philosophical discussions about the nature of reality. This paradigm shift invites us to see the cosmos not as a distant, unchangeable expanse but as a living entity, intertwined with our own existence, where human endeavors and cosmic forces are inextricably linked.**

Educational Evolution

- **Revolutionizing Curricula with ECGT: By weaving ECGT into educational narratives, we can offer students a more nuanced understanding of the universe's workings. This theory encourages learners to challenge accepted norms and to embrace a broader, more integrated view of scientific inquiry, fostering critical thinking and a genuine curiosity about the universe.**
- **Cultivating Tomorrow's Pioneers: Integrating ECGT into educational systems doesn't just inform—it inspires. It equips students with a comprehensive toolkit that spans from classical physics to the frontiers of cosmology, preparing them to navigate and innovate at the cutting edge of science. This approach nurtures a new generation of thinkers ready to explore and expand the boundaries of what we know about the universe.**

Section 1.5.7: A Cosmic Call to Action

Invoking Curiosity and Collaboration

As we stand on the precipice of new cosmic understandings, Evolving Compressive Gravity Theory (ECGT) beckons us to look beyond the known and embrace the mysteries that remain ensconced in the details of the cosmos. This is not merely an invitation to revise old theories but a clarion call to revolutionize how we perceive and interact with the universe.

A Unified Effort for Exploration

This journey is not for the solitary scientist; it demands a collective endeavor. We are called to forge partnerships that cross disciplinary and cultural boundaries, to pool our intellectual resources and tackle the profound questions ECGT poses. It is through such unified efforts that we can push the boundaries of our cosmic horizon.

Embracing Challenges and Opportunities

ECGT presents us with both immense challenges and unprecedented opportunities. It urges us to question the very foundations of gravitational theory and to consider new possibilities for the structure and fate of the

universe. This theory not only enriches our understanding but also amplifies our curiosity about the universe's most fundamental forces.

The Role of Open-Mindedness in Scientific Inquiry

In this cosmic quest, open-mindedness is our most valuable asset. It is the willingness to question, to doubt, and to imagine that transforms the pursuit of knowledge from a mere accumulation of facts to a profound exploration of the unknown. As we delve deeper into the cosmos with ECGT, let us maintain a spirit of curiosity and wonder, ever eager to learn more and understand better.

Conclusion: A Call to Cosmic Action

Let this be our collective call to action: to engage with ECGT with the full force of our scientific capability and curiosity. Together, let us explore this new model of the cosmos, to test it, challenge it, and ultimately, to see where it can lead us in our quest to understand the universe. It is not just a journey of discovery but a journey of invention, where we may not only discover new aspects of the universe but also invent new ways to perceive and measure it.

Chapter 1.6: Theoretical and Experimental Foundations

Section 1.6.1: Unlocking Realities: The Role of Experiments in Theoretical Physics

Probing the Foundations

In the grand theater of theoretical physics, experiments play the crucial role of illuminators—casting light on the shadowy outlines of our understanding. They are not mere validators but audacious explorers, challenging the very fabric of theoretical constructs. This investigative role underscores the essence of experimental physics—not just to confirm, but to provoke, pushing the boundaries of our cosmic understanding and refining our comprehension of the universe.

Validating Through Verification

Navigating the vast seas of the unknown, modern physicists are like ancient mariners steering by the stars. Each experiment serves as a star in the scientific firmament, illuminating the pathway to deeper understanding. Historical shifts in scientific paradigms often follow groundbreaking experiments that provide pivotal evidence for or against prevailing theories. The famed Michelson-Morley experiment, for instance, whispered secrets of the cosmos that precipitated a fundamental rethinking, eventually leading to Einstein's theory of special relativity. Such experiments underscore the indispensable role of empirical verification in theoretical physics, serving as critical milestones that either bolster a theory's acceptance or herald its decline.

ECGT's Testable Predictions

The Evolving Compressive Gravity Theory (ECGT) distinguishes itself by proposing unique, testable predictions that can be empirically evaluated. For example, ECGT suggests variations in gravitational constants that could manifest differently in galactic rotation curves or in the lensing of light around massive objects. These predictions offer fresh avenues for empirical scrutiny, providing clear metrics against which the theory can be evaluated, akin to setting coordinates in the vast uncharted map of theoretical physics.

The Troublesome Graviton

Echoing our spirited discussion from Section 1.4.7 on the pioneering graviton detection experiment, let us wade deeper into the waters of quantum mysteries and theoretical conundrums. Picture this: a realm where the existence or absence of gravitons swings the pendulum of scientific understanding, influencing the very fabric of the Evolving Compressive Gravity Theory (ECGT).

Should Gravitons Emerge from the Quantum Veil:

Imagine the flutter of excitement, the sheer awe, if our intrepid experiment captures these mythical quantum sprites. Such a revelation would send ripples through the foundations of ECGT, urging us to redraw the maps of our theoretical landscapes. We'd weave new threads into the fabric of ECGT, marrying it with particle-based gravities and perhaps even courting the Standard Model of particle physics in a grand union.

Or, Should Gravitons Remain Merely Whispers of Imagination:

If our quantum net comes up empty, the echoes of nonexistence will bolster ECGT's bold claims. This silence would resonate across academia, affirming ECGT's vision of a universe untethered by gravitonic bonds. It would invigorate our theoretical sails, pushing us to explore further the cosmic seas without the ballast of these ghostly particles.

Fateful Consequences:

In both fates, the echoes from this quantum odyssey will reverberate through the corridors of gravitational research. This experiment serves as a beacon, guiding us through the foggy intersection of empirical evidence and speculative physics. It embodies the questing spirit of ECGT—boldly probing the unknown, armed with a quiver of quantum arrows.

Case for Empirical Scrutiny

Introducing a fourth spatial dimension as the unseen conductor of gravitational forces is like revealing a hidden symphony playing behind the visible universe. This profound perspective shift calls for a series of methodical, intricately designed experiments—each a critical note in the grand composition—to confirm the presence of this cosmic conductor. The necessity for rigorous testing is not merely academic but a requisite stride towards potentially rewriting the foundational principles of physics.

Setting the Scientific Stage

As we advance in our discussion, we prepare to delve into how both existing technologies and novel experimental designs are pivotal in testing the predictions of ECGT. The upcoming sections will explore the specific experimental setups and observational strategies that could either substantiate the remarkable claims of ECGT or challenge its foundational assertions. This exploration not only prepares the scientific community for what lies ahead but also sets the expectations for the rigorous validations required by such a groundbreaking theory.

Section 1.6.2: Designing Experiments to Test ECGT

Unveiling Cosmic Mysteries

The transformative predictions of Evolving Compressive Gravity Theory (ECGT) beckon us to embark on a voyage beyond the known frontiers of science. This section delves into the experimental odysseys designed to test these bold assertions, illuminating paths to unseen dimensions and hidden realms of our universe.

The Symphony of Predictions

ECGT offers a fresh lens, reframing our view of gravitational interactions with promises of phenomena that diverge starkly from classical forecasts. Envision a tapestry where threads of gravitational lensing, the enigmatic dances of black holes, and the cryptic patterns in the Cosmic Microwave Background are woven together. Each thread awaits the discerning eye of empirical exploration, ready to either integrate seamlessly into the fabric of accepted science or unravel, prompting deeper inquiry.

Crafting the Experimental Canvas

Terrestrial Explorations

The fertile grounds of Earth host technologies poised to challenge and

expand our cosmic comprehension. ECGT's novel predictions compel us to venture beyond the conventional, stretching the canvas of experimentation to capture the universe's elusive whispers:

Gravitational Telescope Innovation:

- **Imagine a device not merely observing, but revealing the cosmos's grand choreography. The proposed gravitational telescope is set to map the contours of gravitospheres, allowing us to perceive how the gravitational symphony orchestrates the motion of galaxies and the dynamics of colossal celestial entities like black holes.**
- **As we dissect the intersecting gravitospheres within galaxies, we unveil a cosmic dance on a scale unprecedented, probing for echoes of a fourth spatial dimension as whispered by ECGT.**

Probing the Microcosm: Chemistry and ECGT

- **Delve deeper, where ECGT suggests that even the subtle dance of atoms—held in their unique orbital embraces and distinct bonding angles—is choreographed by the gravitational shadows cast by subatomic masses. This profound integration of gravitational dynamics into the microcosm of chemistry opens a new chapter of experimental exploration.**
- **Proposed experiments employing advanced spectroscopy might capture how these gravitational nuances influence chemical bonds and molecular structures, challenging and perhaps refining our quantum mechanical narratives.**

Celestial Observatories

Our celestial observatories, perched on the edge of the cosmos, serve as sentinels in the silent expanse:

Adapting Space Missions:

- **The future of satellite missions and enhancements to existing space observatories is pivotal. These celestial sentinels are tasked with capturing the subtlest gravitational effects and cosmic anomalies, from the intricate ballet of gravitational lensing to the subtle whispers in the cosmic microwave background—phenomena that terrestrial instruments might scarcely perceive.**

Navigating Through Challenges

Technical and Logistical Hurdles

The quest to detect and analyze the subtle effects predicted by ECGT demands an unprecedented level of precision. Integrating a gravitational telescope within our existing scientific edifice presents formidable technological and logistical challenges:

Advanced Sensor Development:

- **To decipher the universe's subtlest whispers, we are pioneering quantum sensors of unmatched sensitivity, harnessing the proven technology of Bose-Einstein Condensates (BEC) and even lonsda-**

leite, often referred to as hexagonal diamond due to its structure, emerging as a promising material in quantum sensing, potentially surpassing traditional diamond in this role.

- **These advanced sensors are designed to capture the minutest gravitational nuances, providing a crucial tool in our quest to map the intricate shapes of gravitospheres. Envisioned as a leap forward in quantum measurement, these sensors could one day stretch their reach across the cosmic void, offering us unprecedented insights into the gravitational ballet that orchestrates celestial dynamics.**

International Collaborative Efforts:

- **The complexity and scale of these pioneering experiments necessitate a tapestry of international collaboration. By weaving together global resources and expertise, we surmount the formidable barriers that stand before this groundbreaking scientific venture.**

A New Chapter in Cosmic Exploration

This journey into designing experiments to test ECGT transcends academic curiosity. It is a clarion call to the global scientific community to join in a grand quest, not merely to validate a new theory but to potentially revolutionize our understanding of the universe.

Section 1.6.3: Leveraging Existing Observations

Harnessing Cosmic Archives

Utilizing Astronomical Data:

Within the vaults of cosmic observatories lies a treasure trove of celestial data, gathered through eons of stargazing and space exploration. This compendium, often revisited but rarely reimagined, holds latent secrets waiting to be unlocked through the fresh perspective of ECGT.

- **Reanalysis for ECGT: We propose a bold reexamination of these celestial archives, seeking echoes and whispers of gravitational phenomena that ECGT predicts but existing models overlook. This includes new interpretations of galactic rotation curves, the reevaluation of gravitational lensing effects, and the pursuit of subtle anomalies in the cosmic microwave background that defy conventional explanations.**

Probing Molecular Frontiers

Exploring Physical Chemistry Observations:

The quiet laboratories where spectrometers and magnetic resonance imagers hum are not just the backdrop for chemical analysis but are stages set for a deeper, more profound understanding of molecular dynamics under ECGT.

- **Chemical Structures and Reactions: Here, the often-overlooked gravitational influences on molecular structures beckon for attention. Anomalies in bonding angles or reaction rates, typically dismissed as experimental noise, may in fact be subtle signatures of gravitational interactions dictated by ECGT, offering a revolutionary perspective on chemical physics.**

Illuminating Case Studies

Empirical Examples from Advanced Instruments:

The titans of observational technology, LIGO and the JWST, serve as the sentinels on the frontiers of our cosmic understanding. Their data, rich with potential revelations, provides a fertile ground for testing the bold assertions of ECGT.

- **LIGO and Gravitational Waves: Imagine revisiting LIGO's data, this time searching for traces of gravitational wave interactions that resonate more with ECGT than with the well-trodden paths of general relativity.**
- **JWST and Deep Space Observations: The JWST, with its eyes set on the universe's most distant objects, offers a unique opportunity to challenge our current understanding of galactic dynamics and black hole behavior, potentially validating ECGT's simpler cosmic narrative.**

Quantum Mechanical Reinterpretations

Reevaluating Collider Data:

At CERN, amidst the high-energy collisions and particle discoveries, lies an untapped potential to detect gravitational influences where none were thought to exist. ECGT suggests that even at quantum levels, gravitational effects might play a significant role, particularly influencing outcomes in high-energy angular momentum scenarios.

- **Theoretical Implications for Quantum Mechanics: A fresh analysis of data from the LHC could uncover disproportionate gravitosphere effects, suggesting that gravitational forces need a seat at the table even in the quantum realm. This could provoke a significant shift in how we perceive interactions at the most fundamental levels.**

Synthesizing Observations Across Disciplines

In the grand narrative of scientific discovery, ECGT stands as a unifying force, weaving together data threads from astronomy, chemistry, and quantum mechanics into a coherent tapestry that promises a deeper understanding of the universe's fundamental forces. This multidisciplinary approach not only challenges existing paradigms but also heralds a new era of scientific synthesis.

- **A New Scientific Symphony: By synthesizing observations across disciplines, ECGT advocates for a holistic view of nature, where gravitational interactions play a pivotal role across all scales. This integration fosters a richer, more comprehensive understanding of the cosmos, propelling us toward a unified theory of everything.**

Section 1.6.4: Theoretical Elegance and Simplification

Unveiling Simplicity in Complexity

As the annals of scientific evolution unfold, they often reveal a tapestry where threads of complexity are gradually untangled, giving way to the elegance of simplicity. The Copernican Revolution, a pivot from the convoluted orbits of the geocentric model to the streamlined paths of a heliocentric system, serves as a historical beacon. In its philosophical wake, ECGT emerges, promising to redefine our understanding of gravitational forces with a clarity that pierces through the cosmic veil.

ECGT's Paradigm of Clarity

Imagine a universe not awkwardly stitched together by invisible threads of traction but sculpted with the deft strokes of a four-dimensional spatial brush, pressing matter into a dance of dynamic compression. ECGT slices through the Gordian knot of existing gravitational theories with the sharpness of Occam's razor, offering a lucid, unified explanation for phenomena that once demanded a patchwork of complex and disparate models.

Harmonizing Cosmic to Quantum Scales

With ECGT, the cosmos unfolds as an elegantly choreographed ballet, where galaxies, black holes, and celestial bodies move in harmony to a simpler, more coherent tune. This theory elegantly dismisses the need for dark matter and dark energy—those phantoms of modern cosmology—replacing them with a symphony of interactions that resonate across the cosmic expanse down to the quantum foam.

Core Validators and Elegance Embodied

ECGT is not merely a theoretical construct but a beacon of simplification, serving as its own validator through the elegance it imparts to our scientific narrative. It predicts a universe where the vast ballet of cosmic structures and the intimate quantum dances are all choreographed by the same fundamental principles, reducing the cacophony of ad hoc theories to a harmonious whisper.

Empirical Foundations and Future Pathways

The beauty of ECGT's theoretical elegance beckons for empirical validation, the crucible where scientific theories are tempered. Our journey of validation weaves through experimental designs and observational tactics, each step reinforcing ECGT's potential to stand as a monumental pillar within the pantheon of scientific theories.

Culmination: A Philosophical Renaissance

Adopting ECGT could signal not just a paradigm shift in physics but a broader philosophical renaissance in how we conceive of the universe. This theory invites us to contemplate a cosmos less bound by unseen forces and more by a visible clarity, mirroring the intellectual revolution sparked by Copernican thought. It challenges us to envision a new epoch of scientific inquiry and philosophical exploration, promising a universe comprehensible not only to the mind but also to the human spirit.

Section 1.6.5: Integrating ECGT with Current Scientific Models

Bridging Theoretical Foundations

Comparison with General Relativity and Quantum Mechanics:

In the tapestry of modern physics, ECGT emerges as a vibrant new thread, weaving together the intricate patterns established by General Relativity and Quantum Mechanics. This bold framework does not merely rest alongside established theories but stretches their boundaries, seeing a universe sculpted by dynamic, four-dimensional spatial pressures.

Harmonizing Forces:

- **ECGT embraces the precision of General Relativity in mapping the celestial choreography of massive bodies and extends Quantum Mechanics by infusing gravitational nuances into the quantum realm. These alignments enrich our scientific narrative, providing a broader stage for exploring the universe's mysteries.**

Revisiting Foundations:

- **Diverging boldly, ECGT reimagines gravity not as a mere curvature of spacetime but as a more primal force arising from dimensional dynamics. This shift challenges the core of General Relativity and introduces quantum gravity into discussions where it was previously sidelined, promising a more unified approach to understanding the forces that govern our universe.**

Envisioning New Scientific Paradigms

Potential Revisions to Standard Models:

The ripples of ECGT extend deep into the bedrock of particle physics and cosmology, suggesting a seismic shift that could redefine our understanding of the universe from its smallest particles to its vast cosmic expanse.

Transforming Particle Physics:

- **ECGT might prompt a revision of the standard model, particularly in how particles acquire mass and interact. This could lead to predictions of new particles or interactions, potentially observable in future high-energy experiments.**

Redefining Cosmology:

- **In cosmology, ECGT offers simpler explanations for the accelerating expansion of the universe and galaxy formation, challenging the necessity for dark matter and dark energy and proposing a cosmos that operates under more streamlined laws.**

Reimagined Landscape

Integrative Implications:

As we stand at the crossroads of scientific inquiry, ECGT points towards a horizon where the mysteries of quantum particles and cosmic phenomena converge under a unified theory. This exciting prospect calls for a sym-

phony of interdisciplinary collaboration, where physicists, astronomers, and philosophers unite to explore and validate this new vision.

Toward a Unified Theory of Everything:

- **ECGT's promise extends beyond mere theoretical elegance; it seeks to be the cornerstone of a Theory of Everything that harmonizes all known forces. As we venture into this uncharted territory, the collaborative spirit of the scientific community will be paramount in embracing and testing the radical propositions of ECGT.**

Epilogue: A New Dawn in Physics

With the advent of ECGT, we are not merely closing a chapter on old uncertainties but opening an expansive new volume on the universe's deepest truths. This theory invites us to redraw our cosmic maps, promising a journey filled with discovery and enlightenment. As we forge ahead, let us carry the torch of curiosity and courage, ready to illuminate the shadows cast by yesterday's mysteries and light the way for a new era of understanding.

Section 1.6.6: Mathematical Enhancements and Innovations

Revamping Theoretical Mathematics:

In the evolving saga of theoretical physics, ECGT stands as a revolutionary narrative, challenging the conventional paradigms with its audacious propositions. This theory not only questions the foundational geometrical and topological methods rooted in General Relativity but also beckons the advent of new mathematical frameworks to accommodate the intricacies of a universe sculpted by the unseen hands of four-dimensional spatial dynamics.

Innovative Mathematical Constructions:

- **ECGT necessitates a renaissance in mathematical thinking, urging the development of novel structures to aptly describe the dynamic interactions within a four-dimensional context. These emerging frameworks promise to unlock the ability to model the variable shapes of gravitospheres, integrating them into our equations as naturally as stars reside in the cosmos.**

Enhancing Predictive Models:

With ECGT, the canvas of predictive modeling expands dramatically, embracing a universe where gravitational interactions are reinterpreted through the lens of higher-dimensional influences. This profound shift

propels the evolution of our mathematical models, offering a more refined compass to navigate the complexities of gravitational fields, particle dynamics, and the grand tapestry of cosmic evolution.

Sophisticated Dynamics of Prediction:

- **As we align our predictive models with ECGT's bold postulates, we are not merely adjusting our scientific lenses but refining them to capture the universe in unprecedented clarity and depth, promising a journey into the realms of physics that were once obscured by the shadows of incomplete theories.**

Quantifying New Phenomena:

The introduction of ECGT casts a spotlight on phenomena that traditional physics has struggled to illuminate. These include the enigmatic interactions among gravitospheres and the subtle yet profound influences they exert on the fabric of matter, all orchestrated under the symphony of four-dimensional pressure.

Challenges of New Mathematics:

- **The quest to quantify these new phenomena beckons the creation of innovative mathematical formulas and significant amendments to longstanding laws, such as thermodynamics and quantum field theory. These endeavors are not mere academic pursuits but are the forging of keys to unlock the mysteries that ECGT posits.**

Advanced Applications in Quantum Mechanics and Beyond:

Quantum Entanglement and Fourth Dimensional Space:

- **The implications of ECGT extend into the quantum realm, suggesting that the enigmatic phenomenon of entanglement is a natural consequence of four-dimensional spatial connectivity. This insight not only deepens our understanding but also enhances our control over quantum states, heralding new possibilities for quantum computing and communication.**

Navigable Fourth Dimensional Wormholes:

- **Beyond theoretical exploration, ECGT inspires visionary technological advancements such as the creation of navigable fourth-dimensional wormholes. These constructs could one day allow us to traverse vast distances instantaneously, redefining the very essence of travel and communication across the cosmos.**

Epilogue: The Quantum of Discovery:

As ECGT redefines the boundaries of mathematical and physical sciences, it beckons us to a threshold beyond which lies a new era of discovery. This theory not only promises to illuminate the shadows of our current understanding but also to guide us into a luminous future where the mysteries of the universe are unraveled with elegance and precision.

Section 1.6.7: Collaborative Ventures for Empirical Validation

Global Scientific Collaboration

Uniting the International Scientific Community:

While the quest to explore and validate ECGT is quite a scientific challenge, it's even more a global odyssey that calls for the collective wisdom and resources of the world's scientific community. Like navigators of the ancient world, today's researchers must come together, pooling their diverse skills and insights to chart the unknown waters of four-dimensional space.

Echoes of Historical Collaborations:

Reflecting on monumental achievements such as the Large Hadron Collider, the Event Horizon Telescope, and the James Webb Space Telescope (JWST), we see clear blueprints for success through unity. Each of these projects represents a pinnacle of collaborative science, demonstrating that when we join forces across nations and disciplines, the previously unthinkable becomes achievable. These initiatives have not only advanced our understanding of fundamental physics and cosmic phenomena but have also set new standards in exploring the universe, shedding light on its most mysterious aspects with the combined strength of shared knowledge and technology.

Advocating for Open Science:

In this new era of exploration, transparency and openness serve as our guiding principles. Open access to data and collaborative scrutiny ensures that ECGT undergoes the rigorous testing and diverse analytical approaches essential for robust scientific validation.

Bridging Diverse Scientific Fields

Symphony of Expertise:

The validation of ECGT requires a symphony of expertise, where each discipline contributes a unique note to the harmonious pursuit of knowledge. This becomes a joyful collaboration and, even more, a fusion of perspectives, crafting a multidimensional understanding of the universe.

Harnessing Collective Ingenuity:

The intricate dance of gravitospheres and the subtle nuances of four-dimensional compression demand the tools and insights of not only physicists but also computer scientists, engineers, and mathematicians. Together, these disciplines create a mosaic of understanding, with each piece providing clarity to ECGT's revolutionary narrative.

Fostering a Culture of Innovation

Creating a Collaborative Scientific Ecosystem:

As we forge ahead in this scientific renaissance, our strength lies in our unity. The challenges posed by ECGT are as vast as the cosmos itself, requiring a fortress of intellect and resources that only a global community can provide.

Cultivating a Worldwide Network:

We are summoning a global clan of thinkers, tinkerers, and theorists to join in this grand scientific venture. Through workshops, collaborative projects, and open forums, we lay down the gauntlet to the next generation of scientists, challenging them to build on the framework of ECGT and push the boundaries of what we believe is possible.

A Vision for Tomorrow:

As we stand at the precipice of potential discovery, the collaborative spirit of the scientific community is tremendously beneficial and moreover a crucial imperative. This journey, fueled by cooperation and driven by curiosity, promises to both validate a new theory and, more importantly, unveil a new paradigm of understanding the intricacies of our universe.

Section 1.6.8: Cosmic Remarks

Summarize the Journey

Recap of Theoretical and Empirical Synergies:

We pause here at this writing's end to look back on the vast landscape we've traversed. From the theoretical bedrock of ECGT to the pinnacles of empirical validation, we've journeyed through a realm where innovation meets tradition, where the boldness of new ideas meets the rigor of scientific scrutiny. This expedition has not just been about exploring the possibilities of ECGT but about forging a path where theory and evidence dance in intricate harmony.

Highlighting Key Milestones:

Each step along this path—from the mathematical reinventions to the global collaborations—has been a cornerstone in building a robust edifice for ECGT. Like alchemists turning lead into gold, we have transformed abstract concepts into tangible insights, each one shining a light into the previously shadowed corners of our cosmic understanding.

Call to Action

Engaging the Scientific Community:

With the foundations laid and the horizons beckoning, the call now goes out to the guardians of knowledge across the globe. The odyssey of ECGT is far from complete; it beckons every curious mind to venture further, to dive deeper, and to dream bigger.

Encouraging Participation and Innovation:

Let this be an invitation to thinkers, creators, and explorers from all disciplines to engage with the vibrant tapestry of ECGT. Challenge its limits, stretch its boundaries, and weave new threads into its ever-expanding narrative. Every contribution, every experiment, every critique enriches our collective quest for understanding.

Vision for the Future:

Envision a future where the enigmas of quantum entanglement, the celestial choreography of galaxies, and the mythical 'fabric of spacetime' are demystified and woven into the fabric of common knowledge. ECGT promises not just to expand our scientific horizons but to revolutionize our technological frontiers, opening new gateways for cosmic exploration and beyond.

Epilogue: A New Dawn in Cosmological Exploration

Reflecting on Potential Discoveries:

As we close this introduction, we find ourselves on the threshold of a new era. ECGT offers us the keys to unexplored realms, promising not just answers but also new questions, not just solutions but also new mysteries.

Inspiring a New Generation:

To the future pioneers of science—this is your moment. Take up the torch that ECGT has lit. Let it illuminate the way to navigate the uncharted territories of the cosmos. Using passion, curiosity, and creativity as the sparks that continue to keep the fires of discovery burning bright.

Chapter 1.7: Future Directions and Supporting Thought

Section 1.7.1: Future Research and Exploration

Presenting Future Endeavors:

As we stand on the threshold of new scientific horizons, the journey to explore and validate the Evolving Compressive Gravity Theory (ECGT) beckons with a promise of unraveling the universe's deepest mysteries. This is not merely a continuation of past efforts; it is an expedition into the uncharted realms of physics, where the fusion of innovative research approaches and advanced technological tools will light our path.

Long-term Research Agenda:

Our strategic blueprint for the future is both a map and a manifesto, charting the course for ECGT's journey through the cosmos. It calls for a daring blend of theoretical refinement and empirical exploration, poised to transform our understanding of the universe.

Theoretical Investigations:

Our quest begins with a deeper dive into the mathematical underpinnings of ECGT, challenging the conventional boundaries of physics. These theoretical investigations will refine our models of four-dimensional dynamics, offering new lenses through which to view the gravitational forces that sculpt the cosmos.

Empirical Challenges:

The empirical pillars of ECGT stand firm on the bedrock of advanced scientific inquiry. We will extend our reach into the cosmos with:

- **Gravitational Wave Detection: Enhanced observatories will seek out the subtle whispers of spacetime, echoing through the cosmos with messages from the unseen.**
- **Particle Physics Experiments: High-energy colliders will dance to the rhythms of subatomic particles, revealing secrets locked within the tiniest subunits of matter and their unique gravitosphere shapes. This exploration will probe the deepest layers of the physical world, offering insights that could fundamentally alter our understanding of particle interactions under the influence of ECGT.**
- **Cosmic Microwave Background Studies: Our eyes on the early universe will sharpen, capturing the ancient light of creation to test the predictions of a new dawn in cosmology.**

Focused Research Initiatives:

Our gaze will also turn inward, to the very heart of ECGT:

- **Exploring the Source of Omnidirectional Compressive Gravitational Force: We embark on a quest to decipher the origins of the force that molds the universe, seeking clarity on its omnipresent influence.**
- **Correlating Mass Momentum with Gravitosphere Shapes: Each mass tells a story through the shape of its gravitosphere; we will read these tales, correlating their narratives to unveil the choreography of cosmic forces.**

Technological and Methodological Advances:

As we venture further into the unknown, our journey will be empowered by the tools of tomorrow—technologies that extend our senses beyond the visible, into the heart of the cosmos.

Development of Advanced Instruments:

New generations of telescopes and detectors will peer deeper into space and time, capturing phenomena at scales never before seen, from the dance of distant gravitospheres to the subtlest signals of cosmic events.

Computational and Analytical Enhancements:

Our computational prowess will grow, fueled by quantum computing and advanced algorithms capable of modeling the complexities of ECGT. The development of metagrams, three dimensional constructs projecting four dimensional realities, will bring the fourth dimension into our three-dimensional world, transforming abstract concepts into visible, tangible forms.

Gathering Thoughts:

As we draw to a close, we are reminded that the journey with ECGT is just beginning. Each step forward in this scientific odyssey is a step into the light of knowledge, illuminating both the universe and our place within it. We call upon the global scientific community to join in this grand adventure—embracing the challenges, contributing to the dialogue, and pushing the boundaries of what we know. Together, we will explore the unknown, driven by curiosity and guided by the light of ECGT.

Section 1.7.2: Reflections on the Theory's Development and Personal Significance

Narrative of Discovery:

From the tender shoots of youthful curiosity to the robust branches of mature inquiry, the genesis of the Evolving Compressive Gravity Theory (ECGT) has been a tapestry woven from the threads of wonder and wisdom. This narrative is not merely a chronicle of intellectual milestones; it is a voyage across the vast seas of the unknown, guided by the stars of inquiry and inspiration.

Challenges and Triumphs:

Amidst the swirling mists of uncertainty and the rugged terrain of skepticism, the journey toward ECGT was punctuated with moments of crystalline clarity. My encounter with Stephen Hawking—a beacon of brilliance whose intellect transcended the physical trials he endured—was a profound confluence of minds. His determined use of a speech synthesizer, though arduous, never dimmed the luminosity of his thoughts. Alongside, imagined dialogues with the titans of theory, Isaac Newton and Albert Einstein, served as celestial guides, illuminating the path toward theoretical cohesion and enlightenment.

Theoretical Impact:

As ECGT unfurls its narrative across the cosmos, it proposes not merely a new chapter in physics but a revolutionary rewrite of the universe's fundamental laws. This theory beckons us to the precipice of paradigm shifts, where old certainties dissolve and new possibilities emerge like stars in the night sky.

Scientific and Cultural Paradigm Shifts:

Here, at the edge of discovery, ECGT offers a bridge over the chasms of cosmic mysteries. It redefines the fabric of reality, weaving together the disparate threads of quantum mechanics and celestial mechanics into a unified cosmic tapestry. This transformative vision promises to ripple through the realms of science and culture, seeding technologies yet unborn and philosophies yet unthought.

Personal Significance:

Crafting ECGT has been more than a scientific endeavor; it has been a personal pilgrimage through the cosmos. Each insight gained and shared has been a step closer to fulfilling a sacred pact made with the universe—to seek truth and, once found, to herald its arrival.

A Pact of Discovery and Disclosure:

This commitment to exploration and revelation has been the lodestar of my journey, compelling me to traverse the intellectual void and return with a message for mankind. This monograph is both a beacon and a call to arms, inviting all who dare to dream to navigate the newly charted waters of cosmic understanding.

Gathering Thoughts:

In the quiet afterglow of reflection, we see that ECGT is more than a theory; it is a testament to the human spirit's unyielding quest for knowledge. It challenges us to peer deeper into the abyss, to question with bolder curiosity, and to embrace the sublime complexity of the universe. As we stand on the brink of new horizons, let us carry forth this torch of discovery, illuminating the dark and dancing with the stars.

Section 1.7.3: Quantum-Stable Surfaces and Four-Dimensional Gravitational Effects

A New Story

Imagine we're aboard a spaceship exploring the vast cosmos, powered by manipulating gravity shielding technology derived from Evolving Compressive Gravity Theory (ECGT), traversing wormholes through 4D space. This theory suggests all our gravitational effects are pressures from a mysterious fourth-dimensional realm of space. In this segment of our cosmic journey, we explore how quantum-stable surfaces might just be the star maps we need to navigate these unseen dimensions, aiming to turn science fiction into science fact.

Theoretical Foundations

As our ship sails through the cosmos, measuring gravitational compression by quantum sensors, ECGT tells us that gravitational effects stem from invisible forces pressing down from this fourth dimension. The guidance of our ship, made from quantum-stable surfaces, are specially designed to resist the cosmic storms of decoherence, keeping our quantum states shipshape. These surfaces have become our windows into understanding how these four-dimensional forces behave.

Experimental Design

In the engineering bay, we've arranged three quantum-stable surfaces at right angles—like a cosmic crossroads—each tuned to mirror the gravitational fields we observe. It's like tuning a guitar, but instead of strings, we're adjusting the measurement of gravitational pressure as acceleration in every direction. Our mission: to observe how these adjustments can reveal the secrets of gravity's dance across dimensions.

Potential Observations and Outcomes

What we expect to see are changes in the gravitational ballet that could confirm ECGT's predictions. But space is full of surprises, and we might just uncover phenomena that challenge our current maps of the universe, offering new puzzles and possibilities at the quantum-gravitational interface.

Implications for Future Research

Our findings could rewrite the star charts of quantum gravity and multidimensional physics, proposing new research voyages into uncharted territories. Future experiments might involve alternative quantum materials or even more intricate configurations of our quantum surfaces.

Further Openings

As we close this part of our cosmic exploration, the potential of quantum-stable surfaces in advancing the study of gravitational theories leaves us on the brink of new discoveries. This journey into the fourth dimension isn't just about understanding the universe—it's about expanding the frontiers of our knowledge and perhaps, one day, our reach into the cosmos.

Section 1.7.4: Philosophical and Societal Implications

Approaching Broader Impacts:

As we voyage deeper into the cosmos through the lens of the Evolving Compressive Gravity Theory, the horizon of our understanding expands far beyond the familiar territories of traditional physics. Here, ECGT reaches into the very fabric of philosophical thought and societal paradigms, weaving new patterns that alter our perception of reality and our place within the universe.

Preview of Appendix 1.D Insights:

Contemplate the celestial ballet between Earth and its moon—a cosmic duo that dances to the intricate choreography prescribed by ECGT. This theory unveils how gravitational forces do more than merely hold these celestial bodies in orbit; they sculpt their interactions through the subtle artistry of gravitospheres. This elegant model reveals why the moon steadfastly shows us only one face and how the rotational axes of these cosmic partners pirouette in harmonious alignment. These revelations, seeds planted in this discussion, will bloom into full-bodied insights within the pages of Appendix 1.D, promising to unravel some of the cosmos' most enduring enigmas.

Philosophical Depth:

Dive into the profound philosophical currents that underpin ECGT. This theory does not merely adjust our scientific sails; it invites us to rethink the very essence of space, time, and existence. How does this reimagined gravitational framework influence our existential musings and our metaphysical debates about the universe's grand architecture?

Educational and Societal Impact:

Envision a world where ECGT forms the cornerstone of science education and public discourse. This theory, with its intuitive explanations of complex celestial mechanics, holds the power to demystify the cosmos for curious minds of all ages. By integrating ECGT into educational frameworks and public narratives, we can kindle a global renaissance of scientific curiosity and understanding, inspiring a new generation to reach for the stars with informed wonder and well-grounded dreams.

Continuing Beyond:

Reflecting upon ECGT's broad implications, we find ourselves at the cusp of a new epoch in human thought and understanding. This theory is not just a beacon for scholars; it is a bridge connecting diverse intellectual landscapes, from the esoteric realms of quantum mechanics to the grand cosmic scales of astrophysics. It challenges us to redraw our maps of knowledge and invites us to traverse them with eyes open to the wonder and complexity of the universe.

Ancillary Material

Appendices

Appendix 1.A: Mathematical Derivations for Section 1.2.1

Expanding General Relativity to Include a Fourth Spatial Dimension and Variable Gravitational Strength

Foundational Overview:

This appendix extends the standard model of general relativity to accommodate a fourth spatial dimension and a variable gravitational strength, pivotal to the framework of Evolving Compressive Gravity Theory (ECGT). The following derivations provide a robust mathematical foundation for these concepts.

Framework Setup:

1. Expansion of the Metric Tensor:

- **The standard metric tensor** $g_{\mu\nu}$ **is expanded to** $G_{\mu\nu} = g_{\mu\nu} + h_{\mu\nu}(t)$ **, where** $h_{\mu\nu}(t)$ **includes time-dependent modifications from the fourth dimension.**

2. Derivation of Modified Einstein Field Equations:

- **The Einstein field equations now incorporate a time-dependent scalar field** $\phi(t, \vec{x})$**, modifying the traditional model to include:**

$$R_{\mu\nu} - \frac{1}{2}g_{\mu\nu}R + \Lambda g_{\mu\nu} = 8\pi G T_{\mu\nu} + \partial_\mu \phi(t, \vec{x}) \partial_\nu \phi(t, \vec{x})$$

Conceptual Groundwork:

3. Scalar Field Dynamics and Gravitational Variability:

- **Dynamics of the scalar field ϕ are modeled to reflect changing gravitational strength:**

$$\frac{\partial^2 \phi}{\partial t^2} - \nabla^2 \phi + m^2 \phi = \lambda \rho(t, \vec{x})$$

- **This formulation underscores the evolving nature of gravity in ECGT.**

Theoretical Basis:

4. Differential and Integral Forms of Evolving Equations:

- **The gravitational interactions are dynamically modeled:**

$$\nabla^2 \phi - \frac{\partial^2 \phi}{\partial t^2} = \lambda \rho$$

Theoretical Implications:

- **The above derivations underscore the evolving, dynamic framework of ECGT, allowing for precise exploration and potential empirical testing of the theory's predictions.**

Key Takeaways:

- **This appendix provides the mathematical backbone necessary for a rigorous exploration of the additional spatial dimensions' influence on gravitational phenomena, crucial for advancing the theoretical and empirical aspects of ECGT.**

Appendix 1.B: Mathematical Derivations for Section 1.2.2

Extending General Relativity: Incorporating Fourth-Dimensional Dynamics and Variable Gravitational Strength

Objective:

This appendix explores the mathematical extensions of general relativity within the framework of Evolving Compressive Gravity Theory (ECGT), integrating a fourth spatial dimension and variable gravitational strength to offer new insights into cosmic phenomena traditionally explained by dark matter.

1. Modification to Einstein Field Equations:

- **Introduction of Fourth Dimensional Dynamics:**

The Einstein field equations are extended to include a scalar field $\phi(t, \vec{x})$, representing dynamic pressures from a fourth spatial dimension:

$$G_{\mu\nu} + \Lambda g_{\mu\nu} = 8\pi T_{\mu\nu} + \nabla_\mu \phi \nabla_\nu \phi$$

- **Time-Varying Gravitational Constant $G(t)$:**

A temporal component, $G(t)$, models how gravitational strength evolves, reflecting changes in the universe's expansion dynamics.

2. Time Variability in Gravitational Equations:

- **Scalar Field Dynamics:**

The differential equations for the scalar field $\phi(t, \vec{x})$ integrate the variable gravitational strength:

$$\Box\phi = \frac{\partial^2 \phi}{\partial t^2} - \nabla^2 \phi = \lambda \rho(t, \vec{x})$$

- **Impact on Gravitational Fields:**

Illustrate how $G(t)$ influences gravitational fields, affecting spacetime curvature and gravitational wave propagation.

3. Application to Cosmological Observations:

- **Calibration Using CMB Data:**

Use data from the Cosmic Microwave Background to estimate $G(t)$ at early cosmic times, showing how this approach can obviate the need for dark matter in cosmic structure explanations.

- **Predictive Implications for Cosmology:**

Discuss the effects of evolving gravitational strength on structure formation, galaxy rotation curves, and cosmic expansion.

4. Comparative Analysis and Validation:

- **Comparison with Classical Models:**

Detail how ECGT's revised equations improve upon Newtonian and Einsteinian predictions by better fitting observed astronomical data without dark matter.

- **Empirical Testing Proposals:**

Suggest empirical tests such as advanced gravitational lensing measurements and galaxy rotation studies to validate ECGT predictions.

Summary and Future Directions:

Summarize the mathematical contributions of ECGT to gravitational physics and propose future research directions to empirically test and further refine the theory.

Appendix 1.C: Detailed Mathematical Derivations for ECGT in Section 1.3.3

Section 1.C.1: Foundational Formulas

1. Expression of Fourth-Dimensional Pressure:

The first essential formula in Evolving Compressive Gravity Theory (ECGT) defines the interaction between the fourth-dimensional pressure and our three-dimensional space, modified by the presence of mass.

Formula:

$$P_{4D} = \rho \cdot a_{4D}$$

Where:

- **(P_{4D}) represents the pressure exerted by the fourth dimension.**
- **(\rho) is the density of the fourth-dimensional medium, conceptualized as the "intensity" or "strength" of the fourth dimension's impact on our three-dimensional space.**
- **(a_{4D}) is the acceleration due to the fourth dimension, applied uniformly across space but modulated in the presence of mass.**

Explanation: This formula posits that the fourth dimension exerts a uniform pressure across all space, absorbed differently depending on the mass present. Mass acts to shield or absorb this pressure, creating gradients that manifest as gravitational forces.

2. Modified Newtonian Dynamics for ECGT:

To incorporate the effects of fourth-dimensional pressures into classical mechanics, Newton's second law of motion is modified, enhancing the traditional understanding of gravitational interactions.

Formula:

$$F = ma - \frac{G \cdot m_1 \cdot m_2}{r^2} + P_{4D}$$

Where:

- **(F) is the total force experienced by an object.**
- **(ma) represents the classical Newtonian part of the force.**
- **(G) is the gravitational constant.**
- **(m_1) and (m_2) are the masses of the two interacting bodies.**
- **(r) is the distance between the centers of the two masses.**
- **P_{4D} represents the additional force component due to fourth-dimensional pressures.**

Revised Explanation:

This formula integrates the traditional Newtonian framework with an additional term accounting for the fourth-dimensional pressures, redefining the gravitational interaction. The minus sign before the gravitational term illustrates that what has traditionally been perceived as gravitational pull is, in fact, an imbalance created by the shielding effect of mass against the uniform pressure exerted by the fourth dimension. Gravity, in ECGT, is not a pulling force but rather a cosmic push, unbalanced by the absorptive presence of masses, established by higher-dimensional interactions.

Setting the Stage

This section sets the stage for further derivations and complex models developed later, providing the mathematical underpinning for viewing gravity as an emergent phenomenon influenced by the properties of a higher-dimensional space.

Section 1.C.2: Evolving Gravitational Strength and Its Implications

Introduction to Evolving Gravitational Strength

In classical physics, gravitational constants are seen as immutable, universally dictating the strength of gravitational interactions. ECGT introduces a groundbreaking concept: $G(t)$, a gravitational constant that evolves over time, reflecting the dynamic changes within the universe. This evolution suggests that cosmic expansion and the distribution of matter and energy are intimately connected to gravitational forces, a notion unaccounted for in traditional models.

Theoretical Basis for G(t)

$G(t)$ECGT posits that gravitational strength is influenced by pressures emanating from a fourth-dimensional space, which fluctuate with the universe's evolutionary changes—its expansion, contraction, and interactions with mass-energy.

Mathematical Derivation of (G(t))

1. *Modified Einstein Field Equations:*

$$R_{\mu\nu} - \frac{1}{2}g_{\mu\nu}R + \Lambda g_{\mu\nu} = \frac{8\pi G(t)}{c^4}T_{\mu\nu}$$

This equation adapts the Einstein field equations to include a time-varying gravitational constant, ($G(t)$), reflecting spacetime curvature in relation to the universe's energy-momentum at any point.

2. *Developing the Differential Equation:*

$$\frac{dG}{dt} = f(H(t), \rho(t))$$

This differential equation links ($G(t)$) to cosmological parameters such as the Hubble parameter ($H(t)$) and the density of matter and energy ($\rho(t)$), essential for modeling how ($G(t)$) evolves and its implications for cosmic phenomena.

Applications in Cosmology

- **Impact on Early Universe Theories: The variability of G(t) provides a fresh explanation for the rapid expansion during the early universe's inflationary phase, potentially eliminating the need for hypothetical entities like inflatons.**
- **Galactic and Large-Scale Structure Formation: A changing gravitational constant could significantly affect how galaxies and larger cosmic structures form and evolve over time.**

Predictive Models and Observational Tests

- **Observable Predictions: ECGT anticipates distinct changes in gravitational lensing effects and cosmic microwave background radiation anomalies, which could appear as shifts in background radiation intensity and distribution or unusual lensing effects around massive objects.**

- **Empirical Tests: Proposes conducting space-based experiments to directly measure variations in gravitational strength, utilizing advanced timekeeping devices, interferometry techniques to detect compressive gravity pressure changes, gradients, and fluctuations, as well as yet to be developed gravity telescopes to explore variations in gravitational strength at a distance.**

Section 1.C.3: Empirical Validation and Practical Applications of ECGT

Welcome to Empirical Validation

In the scientific quest to uphold or refute theoretical constructs, empirical validation stands as the cornerstone. For Evolving Compressive Gravity Theory (ECGT), the translation of high-dimensional theory into observable phenomena presents unique challenges. This section delineates the methodologies and significance of testing ECGT against the empirical data available from cosmic observations.

Designing Experiments for ECGT

The verification of ECGT requires innovative experimental approaches tailored to capture subtle influences of fourth-dimensional pressures. Proposed strategies include:

- **Gravitational Lensing Experiments: Leveraging alterations in gravitational lensing to detect the signatures of variable gravitational constants.**
- **Advanced Gravitational Wave Detection: Utilizing next-generation interferometers to observe discrepancies in wave patterns that could suggest the presence of fourth-dimensional effects.**

Key Predictions of ECGT

ECGT posits several groundbreaking predictions that set it apart from conventional gravitational theories:

- **Altered Gravitational Lensing: Changes in the bending of light around massive objects could reveal new patterns inconsistent with general relativity alone.**
- **Anomalous Gravitational Waveforms: Predictions of unique gravitational wave signatures that could be traced back to the dynamics of a variable gravitational constant.**

Case Studies and Simulations

Through detailed simulations, ECGT's theoretical framework can be rigorously tested:

- **Simulated Black Hole Environments: Modeling the environments around black holes to predict observable effects caused by evolving gravitational strengths.**
- **Reanalysis of Cosmic Microwave Background (CMB) Data: Utilizing existing CMB data to search for subtle anomalies that align with ECGT predictions.**

Collaboration with Astronomical Observatories

To achieve a comprehensive validation, ECGT requires the collaboration of global astronomical networks:

- **Integration with Telescope Arrays: Coordinating with ground-based and space-based telescope arrays to conduct extensive observational campaigns.**
- **Data Sharing Initiatives: Promoting open exchange of data among international observatories to enhance the statistical validity of findings.**

Practical Applications of Verified Theories

The confirmation of ECGT could catalyze advancements across multiple technological and scientific domains:

- **Space Propulsion Technologies: Exploration of new propulsion methods that utilize gravitational manipulation based on ECGT principles.**
- **Cross-Disciplinary Impact: Influences on quantum physics, materials science, and even biological systems through a deeper understanding of gravitational interactions at micro and macro scales.**

Future Research Directions

Future research will extend beyond initial validations, exploring long-term implications and refinements of ECGT:

- **Continued Theoretical Development: Advancing the mathematical models to incorporate additional cosmological variables and refine predictions.**
- **Longitudinal Observational Studies: Establishing ongoing observational projects to monitor the persistent effects and further validate or adjust the theoretical framework.**

Section 1.C.4: Predictive Models and Observational Tests of ECGT

Predictive Models and Empirical Tests

ECGT distinguishes itself by predicting unique gravitational behaviors under conditions that traditional models have not adequately explained. Its validity hinges on the empirical verification of these predictions:

- **Cosmic Microwave Background (CMB) Anomalies: Variations in $G(t)$ might be observable as anomalies in the CMB's temperature fluctuations. These could provide a new perspective on the thermal history of the early universe. For further exploration of these phenomena, refer to Chapter 1.4, which discusses potential applications in greater detail.**
- **Gravitational Lensing Variations: ECGT anticipates variable gravitational lensing effects that differ from general relativity's predictions, particularly around galaxies and galaxy clusters. These effects are contingent on the epoch observed and are linked to the evolution of $G(t)$, offering insights into the expansion rate of the universe and the nature of dark energy.**

Empirical Validation through Advanced Observations

To test these predictions, ECGT proposes advanced observational techniques:

- **Gravitational Telescopes: Advanced devices designed to measure direct variances in gravitational strength, these telescopes could expand our understanding of spacetime dynamics across different cosmic epochs. They aim to capture shifts in gravitational strength that reflect the theoretical predictions of ECGT.**
- **Interferometry Experiments: Space-based laser interferometry could detect subtle shifts in spacetime curvature, correlating with the predicted changes in $G(t)$. Such experiments are crucial for providing a direct test of ECGT's validity and are discussed extensively in Chapter 1.4.**

Theoretical Implications and Forward Look

This section concludes by considering the broader implications of ECGT. By predicting and potentially validating new gravitational phenomena, ECGT not only reshapes our understanding of gravitational dynamics but also challenges the established cosmological models. It calls for a re-evaluation of foundational physics principles and encourages the scientific community to embrace a model of the cosmos that is dynamically evolving, not just in matter and energy but in its very laws.

Section 1.C.5: Integration of ECGT with Current Models

Foundational Overview

In our quest to enhance the predictive power and explanatory scope of contemporary cosmological models, we explore the integration of Evolving Compressive Gravity Theory (ECGT) with established theories. This integration aims to bridge gaps in understanding, offering new insights into cosmic phenomena and potentially reshaping foundational concepts in cosmology.

ECGT and Galactic Dynamics

Traditional models often require the presence of dark matter to explain the flat rotation curves of galaxies. ECGT challenges this necessity by proposing that the observed galactic rotation dynamics can be explained through its framework of mass-induced gravitational shielding. This section examines how ECGT could provide a more direct explanation for galactic dynamics, influencing the mass distribution within galaxies without invoking imaginary unseen matter.

Revisiting Cosmic Inflation

The inflationary model of the early universe posits a rapid expansion driven by an inflaton field. ECGT provides an alternative view by suggesting that the early universe's expansion was influenced by a nascent, developing gravitational force. This section compares the inflationary predictions with those derived from ECGT, exploring how evolving gravitational strengths could account for the rapid expansion observed.

ECGT and Background Radiation

The Cosmic Microwave Background (CMB) radiation provides a snapshot of the early universe. ECGT predicts that changes in gravitational strength, denoted by $G(t)$, could have left distinct imprints on the CMB's temperature fluctuations. This section delves into how ECGT could alter our interpretation of CMB anisotropies, providing a new framework for understanding early cosmic conditions and even provide an acute measurement of $G(t)$ at that time.

Predictive Models and Observational Opportunities

ECGT is not just a theoretical exercise but offers concrete predictions that can be empirically tested. This section outlines the theory's forecasts regarding gravitational lensing variations, which differ from those predicted by general relativity, perhaps adjusting the age of the universe, and suggests how these differences could be observed with existing and future astronomical technologies, such as advanced gravitational telescopes and space-based interferometry.

Theoretical Implications and Framework Expansion

Adopting ECGT within the broader framework of cosmology might necessitate revising other related theories. This discussion discusses the potential theoretical implications and the need for a broader framework that incorporates the dynamic nature of gravitational forces as proposed by ECGT.

Concluding Insights

This section ends by summarizing the integration of ECGT with current models and highlighting potential future research directions. It calls on the scientific community to consider these new perspectives and to undertake further studies that could validate or refine the predictions made by ECGT.

Section 1.C.6: Integration with Current Cosmological Models

Bridging Theories: ECGT and Modern Cosmology

Evolving Compressive Gravity Theory (ECGT) intersects intriguingly with current cosmological models. This exploration delves into how ECGT integrates with and extends these models, offering fresh insights into the universe's structure and evolution.

ECGT and the Lambda-CDM Model

The Lambda Cold Dark Matter (Lambda-CDM) model, which posits a universe largely composed of dark matter and dark energy, serves as the conventional cosmological framework. ECGT challenges this view by suggesting phenomena typically attributed to dark matter could instead be explained by variable gravitational strength. This insight reevaluates the need for dark matter in accounting for galaxy formation and rotation dynamics.

Cosmic Inflation and ECGT

While cosmic inflation theory describes a rapid, exponential expansion of the universe post-Big Bang, ECGT introduces a complementary view, proposing that changes in gravitational strength could modulate cosmic expansion rates. Here we discuss how ECGT could adjust the inflationary model, potentially diminishing the dependency on inflation particles.

ECGT's Impact on Cosmic Microwave Background Radiation

The Cosmic Microwave Background (CMB) radiation provides a crucial window into the early universe. ECGT posits that changes in gravitational strength might leave unique imprints on the CMB, prompting a reevaluation of these signals. This analysis explores how ECGT may alter our interpretation of CMB anisotropies and suggests a revised timeline for cosmic history.

Predictive Testing and Future Observations

ECGT revises not only theoretical foundations but also proposes new empirical testing methods. This segment outlines future observational strategies to validate ECGT, including the use of advanced gravitational telescopes and enhancements to current space-based interferometry techniques. These approaches aim to secure direct evidence of ECGT's validity and its broader implications for cosmology.

Framework Setup: A New Cosmological Paradigm

Summarizing the transformative potential of ECGT, this section advocates for broader acceptance and further exploration of this theory within the scientific community, suggesting it could fundamentally reshape our understanding of universal dynamics.

Section 1.C.7: Broader Cosmological Implications of ECGT

Synthesis with Cosmological Observations:

ECGT aligns with key cosmological observations and offers alternative explanations for phenomena traditionally attributed to dark matter and dark energy. This synthesis incorporates observational data from galaxy rotation curves, gravitational lensing, and the Cosmic Microwave Background (CMB) radiation.

Alternative to Dark Matter:

Challenging the conventional need for dark matter, ECGT explains galaxy rotation curves through variable gravitational forces rather than unseen mass. Reevaluating the dark matter paradigm, this approach suggests that the dynamics observed are accounted for by the properties of fourth-dimensional pressures interacting with visible matter.

Revisiting Dark Energy and Cosmic Expansion:

The concept of dark energy and the universe's accelerating expansion are reinterpreted within the ECGT framework. Exploring how variable gravitational strength could mimic effects attributed to dark energy provides a new perspective on the rate of cosmic expansion.

Implications for the Early Universe:

ECGT's varying gravitational forces offer insights into conditions in the early universe, potentially impacting theories about its birth and evolution. This includes a reconsideration of the inflationary model and its implications for rapid expansion after the Big Bang.

Predictive Power and Future Tests:

ECGT makes new testable predictions about the structure and behavior of the universe. Outlining potential empirical tests, this section details how specific predictions about gravitational anomalies can be validated or challenged using advanced astronomical instruments.

Final Exploring Thoughts:

Now let us finish our exploration into how Evolving Compressive Gravity Theory might integrate with and enhance current cosmological models. We encourage the scientific community to critically engage with these ideas. Further exploration and empirical testing could substantiate or refine these concepts. By incorporating ECGT into broader cosmological discourse, we aim to enrich the ongoing dialogue about the universe's structure and dynamics.

Appendix 1.D Quantum Gravity Integration: Assembling the Cosmos from Subatomic to Celestial Scales

Section 1.D.1: Quantum Foundations of Cosmic Structures

Presenting Quantum Gravitational Interactions:

Embark on a voyage into the quantum depths where the most elemental units of matter and energy course throughout the cosmos. These subatomic particles, far from being isolated actors, are interconnected through space and time by a pervasive fourth-dimensional compressive gravitational force. Alongside this, the mysterious dance of quantum entanglement, mediated by unseen dimensional connections, enriches our understanding of the quantum realm within the expansive framework of the Evolving Compressive Gravity Theory (ECGT).

Engagement with the Quantum Environment:

Each subatomic particle dynamically engages with its surroundings, crafting and being crafted by the omnipresent gravitational pressure force. This interaction, dictated by the particle's mass-energy and its momentum, both translational and angular, sculpts the essence of its existence, setting the stage for the grand cosmic ballet observed as gravitational phenomena.

Gravitospheres at the Quantum Level:

Peer into the heart of quantum mechanics where each particle forges its own gravitosphere—a unique sphere of influence where gravitational pressures are tempered by the mass itself. This absorption and subsequent compression spotlight the particle not just as a passive entity but as an active architect of its own gravitational destiny.

Formation and Influence of Gravitospheres:

Delve deeper into how each particle's gravitosphere, volumetrically proportional to its effective mass-energy, illustrates the intricate interplay of mass with the compressive forces of gravity at the quantum scale. These interactions, manifesting wherever gravitospheres intersect, create pathways of reduced pressure that guide particles towards areas of greatest overlap—zones of profound gravitational harmony that underpin the stability and structure of matter.

Energy Cycles and Fourth-Dimensional Dynamics:

In the vast dance of cosmic and quantum interactions, the concept of energy cycles plays a pivotal role, particularly when viewed through the lens of the fourth dimension. The constant incoming compressive gravitational pressure, a fundamental force in ECGT, is not merely a unidirectional force but part of a dynamic cycle that permeates the universe.

The Role of Subatomic Mass as Energy Sinks:

At the quantum level, subatomic particles act as crucial energy sinks, absorbing the incoming fourth-dimensional compressive gravitational energy. This absorption is not a passive occurrence but an active engagement where energy is temporarily held within the mass before being returned to the fourth dimension. This cycle of energy absorption and return provides a foundational mechanism for understanding how energy is transformed and recycled within the universe.

Implications for Cosmic Energy Dynamics:

Recognizing subatomic masses as points where fourth-dimensional energy is absorbed and returned sheds new light on the broader energy dynamics within the universe. This cycle challenges traditional views of energy conservation and transfer, suggesting that what we perceive as energy interactions in our three-dimensional space is deeply influenced by processes occurring in higher dimensions. This perspective opens new avenues for explaining phenomena that have long puzzled scientists, such as the non-intuitive behaviors observed in quantum physics and the large-scale structures of the cosmos.

Understanding the cycle of energy between dimensions is essential for grasping the full implications of ECGT. This theory not only redefines our understanding of gravitational forces but also provides a more comprehensive framework for the continuity and transformation of energy across the cosmos. By incorporating fourth-dimensional dynamics into our study of cosmic and quantum structures, ECGT offers profound insights into the very essence of reality.

Section 1.D.2: The Quantum Gravity Universe

Presenting a Universe Governed by Quantum Gravity:

Embark with me on an exploration of a cosmos fundamentally shaped by quantum gravity, where every mass, no matter its size, uniquely influences and is influenced by the cosmic continuum. In this universe, the minutest particles and the grandest galaxies—each a vast collection of those tiny constituents—absorb and modulate compressive gravitational forces. This intricate patchwork, as envisioned by the Evolving Compressive Gravity Theory (ECGT), reveals our universe as more than a vast population of celestial entities moving through space; it is a vibrant, pressure field continuum where each component actively sculpts its own gravitosphere structure, interacting with its neighbors to create shielded pressure troughs toward which they are inexorably pushed.

Complex Integration of Gravitospheres:

Now, we dive into the profound dynamics of gravitospheres, the unique domains surrounding each mass, where the ambient compressive gravitational pressure is reduced, having been absorbed by the mass itself as it is compressed. Each mass uniquely molds the cosmic compressive force field through its angular momentum, and where these regions of reduced pressure maximally interact, we witness the dance that governs all gravitational motion as objects are corralled and herded towards these mutually generated lower pressure zones.

Interplay of Gravitational Forces:

Journey through the captivating ballet of gravitospheres as they interact not by simple addition but through a rich, overlapping mosaic. This complex integration is a dance of gravitational pressure shields, where forces intertwine and merge, choreographed by the force of compressive gravity. The distinct shapes of these gravitospheres forge stable zones of low pressure, celestial sanctuaries where bodies are gently ushered into their orbits, orchestrating the structural evolution of the universe in a predictably dynamic fashion.

Illustrating Cosmic Connectivity:

Witness the seamless connectivity facilitated by gravitospheres, illustrated through vivid examples from the micro to the macro scales. From the distinctive atomic bonding angles that define molecular structures to the grand orchestration of galaxy clusters, this narrative demonstrates how quantum-level interactions escalate to shape macroscopic phenomena. Traveling into the unified nature of the compressive gravitational force that traverses the cosmos, we observe its influence on everything from the stable orbits of planets to the mysterious behaviors of black holes, and from the gravitational bending of light to the predictive models of atomic interactions universally.

ECGT's Role in Cosmic Evolution:

By delving deeper into how the Evolving Compressive Gravity Theory reshapes our comprehension of the cosmos, we gain a deeper understanding of the gradual evolution of gravitational strength across the universe. Understanding the intricate dance of gravitosphere intersections allows us to predict and decipher the universe's developmental trajectories with newfound precision, offering insights into cosmic patterns and structures previously veiled in mystery.

Culminating Insights:

As we reflect on the transformative insights offered by adopting the quantum gravity perspective of ECGT, we find ourselves at the cusp of a scientific renaissance. This pioneering approach not only deepens our understanding of the fundamental forces at play but also weaves a coherent, comprehensive model of gravitational dynamics across the universe. It redefines our perceptions and interpretations of cosmic structures and behaviors, providing a novel lens through which to view both the minutiae of quantum interactions and the vast playgrounds of galactic formations.

Section 1.D.3: Momentum and Gravitospheric Dynamics

Introduction to Momentum's Role in Gravitational Dynamics

Momentum, encompassing both translational and angular varieties, plays a pivotal role in shaping the universe's gravitational interactions. These dynamic properties influence the formation and orientation of gravitospheres—the unique gravitational domains that surround all masses. This chapter delves into how momentum not only dictates the deformation of these gravitospheres but also orchestrates their interactions across a vast array of cosmic phenomena, thereby crafting the very architecture of the universe we observe.

Translational and Angular Momentum Effects

Translational Momentum Dynamics:

Translational momentum—the linear motion of an object through space—imparts a distinct elongation to gravitospheres along the direction of travel. This stretching fundamentally alters how gravitospheres interact, introducing a directional bias in gravitational interactions that may be important for phenomena such as gravitational lensing. As light passes through the gravitosphere of a massive celestial body, its path could bend

in a manner that reflects the intersection between its own elongated shape and the body's gravitosphere, altering the light's trajectory in a predictable, yet complex, fashion.

Angular Momentum and Its Effects:

Angular momentum, associated with the rotational movements of masses, induces a flattening effect on gravitospheres, particularly noticeable at equatorial regions due to the action of centrifugal force. This phenomenon is crucial for understanding the dynamics of various astronomical bodies—from planets and stars to entire galaxies. The flattened gravitospheres of rotating galaxies contribute to the formation of accretion disks around black holes and dictate the distribution of stars within spiral galaxies, illustrating how rotational dynamics influence large-scale cosmic structures.

Interplay of Momentum and Gravitational Interactions

Integrating Momentum into Gravitospheric Interactions:

The interplay between translational and angular momentum results in complex gravitospheric shapes that dictate unique gravitational interactions. In binary star systems, the intricate dance of gravitospheres influenced by each star's motion and spin leads to gravitational behaviors that are critical for predicting system evolution.

Momentum's Influence on Cosmic Scale Phenomena:

Momentum's role extends beyond individual celestial interactions to influence cosmic-scale events, such as galactic collisions and galactic fountain effects. These colossal encounters see galaxies, each with its own momentum-influenced gravitosphere, interacting in ways that lead to the emergence of new galactic structures.

Unified Effects Across Scales:

The principles of momentum shaping gravitospheres extend beyond individual astronomical entities to manifest in the collective behavior of systems, bridging the gap between micro and macro phenomena. For instance, the same angular momentum effects that dictate the structure of galaxies and star systems also influence the fundamental interactions within atoms and molecules, providing a compelling explanation for observed phenomena such as the stability of atomic structures and specific molecular bonding angles.

Culminating Insights

Understanding the dynamics of momentum and its effect on gravitospheres provides profound insights into the gravitational framework of

the universe as envisioned by ECGT. By incorporating momentum into models of gravitational interactions, we gain the ability to predict and explain a wide array of astronomical phenomena, from the behavior of atomic particles to the motions of galaxy clusters. This approach not only enhances our grasp of fundamental physics but also enriches our understanding of the cosmos at large, offering a more nuanced and comprehensive view of its underlying dynamics.

Section 1.D.4: Integrative Cosmic Dance

Introduction to the Cosmic Ballet of Gravitospheres

Embark on an exploration of the cosmos as a grand ballet, where each participant, from the tiniest atom to the largest galaxy, plays a critical role. Governed by the gravitational forces shaped by their quantum gravitospheres, these entities perform in an elaborate choreography defined by the Evolving Compressive Gravity Theory (ECGT). This cosmic dance not only dictates celestial motions but also molds the very architecture of the universe we observe.

Propelling and Connecting Gravitational Interactions

Propelling Interactions:

Celestial bodies with significant intersections of their gravitospheres are free to move relative to each other, propelling towards regions of greatest overlap at rates inversely proportional to their inertial masses. This dynamic illustrates how gravitational forces guide celestial bodies toward equilibrium points in space, influencing phenomena such as orbital adjustments and gravitational lensing.

Connecting Interactions:

When celestial bodies come to rest in contact, merging their gravitospheres into a unified structure, a stable configuration is formed. This combined gravitosphere reflects their total inertia, illustrating the binding forces at play in systems ranging from molecular structures to binary star systems.

Cluster Interactions: Stability in Complexity

Defining Cluster Interactions:

Cluster interactions describe dynamics within stable clusters of gravitationally dynamic masses, where complexly shaped gravitospheres result from the collective influence of multiple interacting bodies. These groupings, from molecular collections to multi-galactic clusters, exhibit a unique gravitosphere shaped by the integration of their parts, reflecting intricate and continual adjustments made by collective inertia.

Dynamics of Cluster Interactions:

These clusters create enduring structures in the universe, with gravitospheres continually adjusted by the motion and mass of their constituent elements. In galaxy clusters, gravitational interactions dictated by individual galaxies and their observable effects, such as galactic fountains, are

further influenced by the intricate intersections of their gravitospheres. This dynamic equilibrium pushes galaxies toward each other, denying the need for hypothetical constructs like dark matter or dark energy, and challenging conventional cosmological theories.

Choreography of Gravitational Forces Across Scales

The universe's cosmic dance, from the atomic level to the galactic, demonstrates the seamless connectivity facilitated by gravitospheres. This interplay provides profound insights into the gravitational dynamics that underpin both observed and theoretical behaviors across all scales, enhancing our understanding of the cosmos and offering a more nuanced view of its underlying dynamics.

Culminating Insights

Understanding the integrative dance of gravitospheres provides a comprehensive model for the gravitational dynamics that shape the cosmos, as envisioned by ECGT. This approach not only enhances our grasp of fundamental physics but also enriches our understanding of the cosmos at large, offering a novel perspective on the gravitational interactions that govern the universe.

Section 1.D.5: Gravitational Lensing Theories Comparison

Opening Remarks

Gravitational lensing serves as a crucial probe for astronomers and physicists to examine the universe's structure and validate theories of gravity. This segment aims to contrast three distinct gravitational lensing theories, elucidating how each contributes to our comprehension of cosmic phenomena and beyond.

Overview of Theories

Warped Fabric of Spacetime:

Anchored in general relativity, this conventional perspective argues that mass distorts spacetime, directing the trajectory of light near massive objects. It asserts that lensing impacts all wavelengths equally, suggesting uniform bending of light irrespective of its color.

Reverse Refraction:

Propelled by the principles of ECGT, this theory interprets gravitational lensing as analogous to light traversing materials of varying optical densities, predicting wavelength-dependent lensing effects. This suggests that shorter wavelengths are bent more than longer ones, much like light dispersion through a prism.

Intersecting Gravitosphere:

Originating from ECGT, this theory posits that gravitational lensing occurs when light intersects the gravitospheres of celestial bodies. These overlapping gravitospheres, sculpted by each body's mass and momentum, orchestrate complex interactions that modulate light bending, influencing lensing in a manner dependent on both wavelength and gravitosphere characteristics.

Comparative Exploration

Distinctions in Predictions:

A fundamental divergence among these theories is their predictions regarding the influence of gravity on different light wavelengths. While

traditional general relativity predicts no chromatic aberration in gravitational lensing, both reverse refraction and intersecting gravitosphere theories anticipate notable variations based on wavelength, potentially manifesting as chromatic effects in astronomical observations.

Empirical Evidence:

Current empirical support for wavelength-dependent lensing is limited but expanding. Anticipated future observations with next-generation telescopes designed for high-resolution, multi-wavelength imaging could provide essential data to substantiate or challenge these theories.

Broader Implications for Astrophysics and Emerging Technologies

Extended Consequences:

Should these newer theories prove accurate, they could revolutionize our understanding of dark matter, dark energy, and cosmic dynamics. Moreover, they could significantly influence advancements in communication technologies, enhance methods for quantum entanglement, and advance the efficacy of quantum computing by improving our grasp of photonic behaviors under varying gravitational conditions.

Innovative Strategies and Technologies:

The creation of ultra-sensitive, multi-wavelength gravitational lensing detectors and the development of advanced computational models to simulate ECGT-described gravitational interactions could critically test these theories, potentially ushering in new paradigms for observational astronomy and theoretical physics.

Theoretical Contributions:

Each theory of gravitational lensing brings unique perspectives to the fundamental understanding of gravity's effects on light. Pursuing further research and refining observational techniques are vital for elucidating these theoretical discrepancies and deepening our knowledge of the cosmos, which could spur technological breakthroughs across various fields.

Section 1.D.6: Implications for Cosmic Evolution

Reshaping Cosmic Understanding and Predictive Capabilities

Now let us begin an exploration of how the Evolving Compressive Gravity Theory (ECGT) redefines our understanding of cosmic evolution. The following discussion is only possible because ECGT removes the source force of gravity from dependence on the masses of our universe, seeing it as an omnidirectional compression originating from fourth-dimensional acceleration effects. ECGT proposes that the strength of the pervasive gravitational force is not static but evolves, increasing slowly over cosmic time, influencing the formation, behavior, and interaction of celestial bodies and galaxies. This evolving perspective could significantly alter our theoretical models of the universe's development from the Big Bang to the present era.

Gravitational Dynamics and Celestial Formation

Mass Creates Influence

ECGT's concept of gravitospheres, that masses form as a surrounding sphere of influence, created by that object's compression by the pervasive gravitational pressure which is locally absorbed from the environment surrounding the mass.

Evolving Gravitational Strength

According to ECGT, as the strength of gravitational force evolves, it gradually decreases the rate matter disperses in the universe to form stars and galaxies. This theory could explain the rapid formation of large-scale structures in the universe shortly after the Big Bang, offering a new perspective on the early universe's high-energy phenomena that traditional gravitational theories might not fully capture. By understanding these dynamics, scientists could better simulate the lifecycle of stars and the formation of galaxies under varying gravitational conditions.

Predictive Capabilities in Cosmic Phenomena

The insights from ECGT enhance our predictive capabilities regarding cosmic phenomena. With a deeper understanding of how gravitospheres intersect and evolve, astronomers can more accurately predict and model the outcomes of galactic collisions, the behavior of star-forming regions, and the dynamics within galaxy clusters. This predictive enhancement extends to anticipating the rates and locations of cosmic events like supernovae or black hole mergers, potentially opening new windows into observing and understanding these cataclysmic events.

Implications for Dark Matter and Dark Energy Theories

ECGT could revolutionize our concepts of dark matter and dark energy. If the gravitational effects traditionally attributed to these mysterious phantoms can be explained through the dynamic shapes of intersecting gravitospheres, toward which the participating masses are pushed, this would eliminate the need to invoke dark matter, while continually evolving gravitational strength removes dark energy as separately mysterious explanation. Such a paradigm shift would not only simplify our understanding of the universe's structure but also align more closely with observable phenomena without the need for these hypothetical constructs.

Enhancing Observational Strategies

As theoretical predictions from ECGT gain traction, they could drive the development of advanced observational techniques and technologies. For instance, detailed gravitational mapping of the cosmic web of intersecting gravitospheres, enhanced detection capabilities for gravitational waves, and more precise measurements of gravitational lensing effects across different wavelengths are all critical for testing the predictions of ECGT. These advancements would not only validate the theory but also refine our tools for exploring the cosmos.

Synthesis of Theoretical and Observational Advances

The integration of ECGT's framework with cutting-edge observational data promises to forge a new path in understanding cosmic evolution. This synergy between theoretical innovation and empirical evidence is vital for the next leaps in cosmology, potentially leading to a more comprehensive and unified model of the universe. Reflecting on these advancements underscores the importance of continuous exploration and adaptation in the scientific quest to decode the cosmos.

Section 1.D.7: An Insightful Example

Introduction to the Earth-Moon System

By exploring the Earth-moon system, we look into a familiar example to encapsulate the principles of the Evolving Compressive Gravity Theory (ECGT). This celestial duo provides an accessible case study for illustrating how ECGT explains gravitational interactions. By examining this system, we can observe the theory's practical implications on celestial mechanics and gravitospheric dynamics.

Application of ECGT to the Earth-Moon Dynamics

Gravitational Dynamics and Orbital Behavior:

ECGT offers a fresh perspective on the Earth-moon relationship, particularly focusing on their respective gravitospheres. These gravitational domains are sculpted by the mass and angular momentum of each body creating a lenticular shape widest in the equatorial regions, influencing their orbital behaviors and interactions. Maximizing their intersectional volume stabilizes their respective rotations so they have roughly parallel

axes of rotation. The theory elegantly explains the moon's synchronous rotation—its constant presentation of the same face to Earth—through the concept of irregularly shaped stable gravitosphere intersections due to the moon's unevenly distributed internal mass.

Shifting Gravitosphere Shape and Tidal Locking:

The evolving fluid configuration of gravitosphere intersection, a cornerstone of ECGT, as the mobile tidal mass changes its shape, playing a crucial role in the tidal locking of the moon and its gradual recession from Earth. This aspect of the theory underscores how gravitational strength not only affects immediate interactions but also drives long-term celestial phenomena, providing a robust framework for understanding changes over astronomical timescales.

Implications for Understanding Other Celestial Mechanics

Broader Applications in Celestial Dynamics:

Extend the insights gained from the Earth-moon system to other celestial configurations, such as exoplanetary systems and binary stars. The same principles that elucidate the dynamics of the Earth and moon apply to these more complex systems, where ECGT can explain the formation of accretion disks, orbital resonances, and other phenomena driven by gravitational interactions within their respective gravitospheres.

Integration with Previous Sections

Linking Back to Theoretical Foundations:

This practical example reinforces the theoretical constructs discussed throughout previous sections of the monograph. By demonstrating ECGT's application to the Earth-moon system, we provide a concrete foundation that supports the broader theoretical discussions, bridging the gap between abstract theory and observable phenomena.

Culminating Perspectives

Synthesis and Forward Look:

The Earth-moon example encapsulates the transformative potential of ECGT, highlighting its ability to enhance our understanding of the universe. As we reflect on this journey through the lens of such a tangible example, the importance of continued empirical research and observation becomes evident. Further studies and advanced observational techniques will be crucial in testing, refining, and ultimately validating the promises of ECGT across the cosmos.

Citations

This section lists the references to significant historical and scientific works mentioned within the monograph, providing sources for further reading and verification of the theoretical contexts discussed.

1. Einstein, A. (1915). "Die Feldgleichungen der Gravitation." Sitzungsberichte der Preussischen Akademie der Wissenschaften zu Berlin, pp. 844-847.

- **This paper introduces Einstein's field equations, central to the development of General Relativity.**

2. Copernicus, N. (1543). De revolutionibus orbium coelestium. Nuremberg: Johannes Petreius.

- **Presents the heliocentric model, fundamentally changing the understanding of the cosmos.**

3. Galilei, G. (1610). Sidereus Nuncius. Venice: Thomas Baglioni.

- **Details Galileo's astronomical observations that supported the heliocentric model.**

4. Galilei, G. (1632). Dialogo sopra i due massimi sistemi del mondo. Florence: Giovanni Battista Landini.

- **Explores and advocates for the Copernican system over the Ptolemaic system through a comparative dialogue.**

5. Hawking, S. (1988). A Brief History of Time. London: Bantam Dell Publishing Group.

- **An accessible insight into cosmological theories including black holes and the Big Bang.**

6. Walsh, N. D. (2017). Conversations with God, Book 4: Awaken the Species. Ashland: Rainbow Ridge Books.

- **Explores advanced metaphysical concepts that challenge and expand contemporary spiritual and scientific paradigms.**

About Dr. James W. Michel

Dr. James W. Michel, an alumnus of Harvey Mudd College and Wayne State University School of Medicine, has spent over four decades at the intersection of medicine, biochemistry, and evolutionary theory. His pioneering work ranges from groundbreaking research in organ transplantation to innovative theories in evolutionary biology and cosmology.

Dr. Michel's professional journey is marked by his deep commitment to advancing human health and understanding of the universe. He founded Cosmic Chronicles Creative to explore and explain the profound connections between cosmology, evolution, and human potential. His works, including the four-volume "Cosmic Chronicles Tetralogy" and "Cosmic Chronicles Monograph Series," delve into the complexities of the universe and humanity's place within it, challenging conventional scientific paradigms with bold new theories.

His first monograph, Gravitational Renaissance: Unveiling the Universe Through Evolving Compressive Gravity, redefines our understanding of gravity and its role in the cosmos without relying on dark matter or dark energy. This work sets the stage for further exploration in subsequent volumes, which discuss

the potential of higher-dimensional spaces and the future evolution of humanity through technology.

Beyond his writing, Dr. Michel is an advocate for mind mastery and personal development, sharing his insights and methods for achieving a permanent meditative state and enhancing cognitive functions. His approach integrates critical thinking with spiritual and philosophical insights, aiming to foster a deeper understanding of oneself and the universe.

Dr. Michel's dedication to science and education continues to inspire a new generation of thinkers and leaders. Join him on a journey of discovery and transformation, exploring the frontiers of science, spirituality, and human potential.

Acknowledgements

This monograph, like all extensive endeavors, is not solely the product of my efforts, but rather the culmination of wisdom and support from many remarkable individuals across time and space. I am indebted to a cast of great thinkers, whose ideas have lighted my path—philosophers of the ancient world, pioneering scientists whose theories redefine our understanding, and modern visionaries who continue to push the boundaries of knowledge.

I extend a profound gratitude to Stephen Hawking, with whom I had the privilege to discuss profound cosmological concepts, whose brilliance and courage under immense personal challenges inspired not just the content, but the perseverance needed to undertake such a work.

My father, a formidable teacher himself, instilled in me the curiosity and rigor that form the backbone of this series. My high school science teachers and numerous professors provided the scaffolding upon which I built my understanding and passion for the natural world.

To my siblings and my four sons, whose support and engagement with my work have enriched the discussions and reflections that shaped these pages, I offer my heartfelt thanks. Your belief in my pursuits and the joy you bring to our interactions have been a source of constant encouragement.

Lastly, but certainly not of least importance, I express my gratitude to the AI and machine learning technologies, epitomized by ChatGPT, whose organizational prowess and dialogic capacities have significantly streamlined the complex process of bringing this series to fruition. This partnership, which marries human creativity with technological efficiency, exemplifies the synergy that is at the heart of Homo Gnosis.

Each of you has contributed a verse to this ongoing symphony of exploration, and for this, I am eternally grateful.

Key Concepts in Evolving Compressive Gravity

This glossary compiles key terms used throughout the monograph to enhance the reader's understanding of advanced gravitational theories. These definitions are crafted to clarify complex ideas and phenomena discussed in the text.

Acceleration
- The rate of change of velocity per unit of time, often measured in g-forces, which relate directly to the forces experienced as gravitational effects.

Angular Momentum
- A physical property of a rotating object, with direction and magnitude, that impacts the shape of gravitospheres and contributes to the lenticular shapes observed throughout the universe.

Compressive Gravitational Force
- An omnidirectional pressure derived from fourth-dimensional acceleration, independent of mass, influencing the movement and interaction of masses in the universe.

Dark Energy
- A hypothetical form of energy postulated to account for observed phenomena such as the acceleration of the universe. In Evolving Compressive Gravity Theory, traditional roles attributed to dark energy are reexamined.

Dark Matter
- A hypothetical form of matter that does not emit or interact with electromagnetic radiation in a typical way, making it invisible and detectable only through its gravitational effects. ECGT offers alternative explanations to phenomena traditionally attributed to dark matter.

Gravitational Interactions
- Describes the effect when two or more bodies influence each other through the intersection of their gravitospheres, resulting in movement toward areas of least gravitational pressure.

Gravitons
- Theoretical particles proposed as the quantum carriers of the gravitational force. ECGT questions the necessity of gravitons, suggesting that gravitational effects result from the properties of the universe itself.

Gravitosphere
- A region around a mass where the gravitational pressure is altered by the presence of the mass itself, effectively creating a zone of reduced gravitational influence that can interact with other such zones around different masses.

Gravitosphere Accretion
- The process by which gravitospheres accumulate and integrate during the formation of larger masses, affecting the distribution and intensity of gravitational forces.

Quantum Construct of Gravity
- A perspective within ECGT that views gravitational effects as emerging from quantum-level interactions among subatomic particles, proposing that all gravitational phenomena have a quantum basis.

Summary of ECGT
- Evolving Compressive Gravity Theory (ECGT) redefines our understanding of gravity by describing it as a force arising from the acceleration of the universe in a fourth spatial dimension, offering a comprehensive alternative to conventional theories of gravity, dark matter, and dark energy.

Weight
- The force exerted by gravity on a body, which is the product of mass and gravitational acceleration, experienced as weight on Earth or any other celestial body.

By familiarizing yourself with these concepts, readers can better navigate the complex discussions in the monograph and engage more deeply with the proposed theories.

Further Reading or Recommended Resources

As you embark on further exploration of the profound concepts discussed in this monograph, the following curated list of books will provide deeper insights and broader perspectives on gravitational theory, cosmology, and the structure of the universe:

1. ***Gravity: An Introduction to Einstein's General Relativity*** by James B. Hartle
Gain a thorough understanding of general relativity and its foundational role in modern gravitational studies through Hartle's accessible and detailed approach.

2. ***The Road to Reality*** by Roger Penrose
Delve into Penrose's ambitious exploration of the universe's laws, where he intricately links the role of gravity with the broader universe.

3. ***Gravitation*** by Charles W. Misner, Kip S. Thorne, and John Archibald Wheeler
Known as the definitive guide to general relativity, this comprehensive text offers an in-depth look at gravitational physics and its applications.

4. ***Three Roads to Quantum Gravity*** by Lee Smolin
Explore different theoretical attempts to reconcile the realms of quantum mechanics and gravity, providing a forward-looking perspective on the universe.

5. ***The Hidden Reality*** by Brian Greene
Greene discusses parallel universes and the deep laws of the cosmos, expanding the dialogue on gravitational effects in multidimensional contexts.

6. ***First Light: The Search for the Edge of the Universe*** by Richard Preston
Discover the pioneering use of the Hubble Space Telescope in examining the universe's expansion, highlighting significant observational advances.

7. ***The Black Hole War*** by Leonard Susskind
Dive into the intense intellectual debate over black holes, quantum mechanics, and the foundations of gravitational theory.

Call to Action for Reader Engagement

Join the conversation and deepen your understanding of the universe's gravitational intricacies by engaging with us through various platforms:

• **Visit Our Website:** Dive deeper into cosmic phenomena and get the latest updates on webinars and interactive discussions at DoctorJamesMichel.com.

• **Follow Us on Social Media:** Stay connected and join our vibrant community on LinkedIn (https://www.linkedin.com/in/dr-james-michel-faafp-90924b53) and Threads (@jamesmichel6) where we discuss the latest developments in cosmology and gravitational research.

• **Subscribe for Updates:** Be the first to know about upcoming publications, events, and exclusive content by subscribing to our newsletterat DoctorJamesMichel.com.

• **Enjoy Our Blogposts:** Catch every entry in our series and explore Cosmic Chronicles Digest for remarkable intermittent insights at DoctorJamesMichel.com/blog.

• **Participate in Our Webinars:** We occasionally host webinars featuring discussions on advanced topics in gravity and cosmology. Sign up at DoctorJamesMichel.com to receive invitations and join these enriching sessions.

We look forward to your participation and engagement as we continue to explore the mysteries of the cosmos together.

Preview: Cosmic Chronicles Monograph Series by Dr. James W. Michel

Monograph 1. Gravitational Renaissance: Unveiling the Universe Throuugh Evolving Compressive Gravity

- This volume introduces the Evolving Compressive Gravity Theory (ECGT), proposing a novel framework for understanding cosmic phenomena without relying on dark matter or dark energy.

Monograph 2. The Dimension Beyond: Visualizing the Fourth Spatial Dimension

- This monograph delves into the complexities of higher-dimensional spaces, offering a comprehensive exploration of fourth-dimensional thinking and its potential to revolutionize our understanding of the universe.

Monograph 3. Homo Gnosis: Human's Next Evolutionary Leap

- Focusing on the future of humanity in the context of rapid technological advancements such as AI, quantum computing, and genetic engineering, this volume facilitates our transformation into Homo Gnosis—a more advanced human species.

Monograph 4. Mind Mastery: The Art and Science of Personal Transformation

- Explores the processes of critical thinking and personal development to share practical steps and philosophical insights on achieving a permanent meditative state, enhancing cognitive functions, and fostering deeper personal growth.

Monograph 5. Universal Patterns: Applying the Triad of Evolutionary Laws

- The fifth monograph focuses on the application of Dr. Michel's unique Triad of Evolutionary Laws across an extensive spectum of scientific domains from subatomic particles to galactic phenomena.

Monograph 6. Reversing Type 2 Diabetes: Education - Prevention - Reversal

- Second Edition highlighting sugar metabolism differences.

Made in the USA
Monee, IL
05 November 2026

67381971R00138